A PLANT HUNTER IN TIBET

A MISHMI GRAVE, DECORATED WITH TIBETAN PRAYER FLAGS

A PLANT HUNTER IN TIBET

F. Kingdon-Ward
F.R.G.S. [Gold Medallist]

Orchid Press

Frank Kingdon-Ward
A PLANT HUNTER IN TIBET

First published by Jonathan Cape Ltd., London, 1934
Second edition, 2006, 2019

ORCHID PRESS
P.O. Box 19,
Yuttitham Post Office,
Bangkok, 10907 Thailand
www.orchidbooks.com

Copyright © Estate of Frank Kingdon-Ward. Protected by copyright under the terms of the International Copyright Union: all rights reserved. No part of this publication may be reproduced in any form or by any means, electronic or mechanical, including photocopying, recording, or by any information storage or retrieval system without prior permission in writing from the publisher.

ISBN 978-974-524-087-2

Dedicated to
BILLY CROFTON

(Dr. W. M. Crofton, M.D.)

who gave me back the strength to climb mountains

CONTENTS

	PREFACE	ix
	INTRODUCTION	xi
I	UNADMINISTERED TERRITORY: JUNGLE WAYS	1
II	THE HOT VALLEY	11
III	THE VALLEY WHERE NO MAN GOES	20
IV	FIRST FLOWERS	30
V	ORDEAL BY FLOGGING	37
VI	THE RIVER OF ICE	48
VII	PLANT HUNTING	58
VIII	FILMING THE FLOWERS	67
IX	OVER THE GREAT SNOW RANGE	73
X	THE LONE MONASTERY	80
XI	HIGH LIFE ON THE ROOF OF THE WORLD	88
XII	EXPLORING	97
XIII	THE RIDDLE OF THE SALWEEN	107
XIV	NINGRI TANGOR	115
XV	THE DESERTED FORT	124
XVI	THE FIGHT IN THE DZONG	130
XVII	LOST ON THE GREAT RANGE	137
XVIII	THE ROAD TO RU	145
XIX	FAREWELL TO THE MONASTERY	151
XX	THE PILGRIM'S PROGRESS	161
XXI	THE GLAMOUR OF THE MOUNTAINS	169
XXII	OVER THE LAST PASS	177
XXIII	THE MISHMI HILLS	185
	NOTES	193
	INDEX	201
	ABOUT THE AUTHOR	211

ILLUSTRATIONS

A MISHMI GRAVE, DECORATED WITH TIBETAN PRAYER FLAGS	*Frontispiece*
A KAMPA WOMAN	17
MAGNOLIA CAMPBELLII	27
IRIS WATTII	27
GECHI GOMPA	39
THE WOOD TURNERS	55
MY COOLIES ABOVE THE ATA GLACIER	63
CHORTENS AT SHUGDEN GOMPA	77
SHUGDEN GOMPA	83
GLACIERS ON THE GREAT SNOW RANGE	94
RAMBU GOMPA	101
THE LAKE, SHUGDEN GOMPA	117
A MENDICANT MONK	127
THE ABBOT OF SHUGDEN GOMPA	127
GENTIANS AT SHUGDEN GOMPA	141
STELLERA CHAMAEJASME	141
DORJE TSENGEN	153
A KAMPA GIRL	163
ON THE JARA LA: THE ASSAM MOUNTAINS AND SOURCES OF THE DIBANG	181
MAP I- SHOWING AUTHOR'S ROUTES	44-45
MAP II - THE HIMALAYA EAST OF THE TSANGPO	208-209

PREFACE

To all those who assisted me on my expedition, my thanks are due. I am specially indebted to several friends of long standing: to Captain John Ramsbottom, O B E., Keeper of the Botanical Department, Natural History Museum; and to Mr. T. P. M. O'Callaghan, Inspector-General of Police, Assam. I would also wish to thank Lt.-Colonel H. W. Tobin, D.S.O., and Mr. J. H. Crace, Political Officer, Sadiya, for assistance on the spot.

This book is a small tribute to the company of enthusiastic gardeners who supported and encouraged me, particularly to my friends of the Garden Society.

F. K. W.

Hatton Gore,
   Harlington,
      Middlesex.

INTRODUCTION

ALTHOUGH the surface features of the high Asian plateau called Tibet are now known, there is no region in the world of comparable size which is so unknown in detail. Even the surface features have only been revealed during the present century, and are familiar to but few; and there are still considerable gaps. There are no maps of Tibet on a scale larger than sixteen miles to an inch, and these show many blank areas, and still greater areas filled in between the tracks of explorers in which the mountains, lakes and rivers are mainly conjectural.

If this is true of the main surface features, naturally the first objects to be noted by the pioneer traveller, how much more true is it of secondary features—the rocks and their structure, animal life and vegetation! It may be argued that Tibet being, as is commonly supposed, a cold desert, the vegetation at least is so sparse that it is of very little importance, and that our knowledge of the flora of Tibet is complete. Tibetan explorers in fact, for the last century, have been so surprised to meet with any plants at all, that they have invariably collected what they saw. In the almost complete absence of vegetation, it is not to be supposed that there can be much animal life either, so that the fauna also might be well known. This is true so far as it goes; but it is not the whole truth. Had Tibet been known to the ancients as Arabia was, it would have been divided into three parts. These might have passed down to history as *Scythia borealis, Scythia felix* and *Scythia horridula*; or, as we would say, the northern plateau, the outer plateau and the river gorge country. We will now briefly consider how these three divisions of Tibet have arisen. Tibet, and the Himalayan ranges, at the end of Cretaceous times lay at the bottom of the great sea known as the Tethys which stretched away to the Mediterranean. To the south lay peninsular India, which had been land almost from the beginning of geological time; to the south-east lay the ancient Indo-Malayan mountains, where now Burma, Yunnan and Malaya are. During early and middle Tertiary times, a series of great earth movements uplifted the bottom of the Tethys into the plateau of

INTRODUCTION

Tibet, and at the same time raised the southern edge of this plateau into the greatest mountain range in the world—the Himalaya. Thus Tibet is, geologically speaking, a new land. Ever since it rose from the sea the forces of nature have been violently destroying and re-shaping it. As a result, we can distinguish three divisions of the plateau:

(i) The Chang Tang, or great plateau. Drainage internal, short rivers ending in lakes which, having no outlet, are salt or brackish. The climate is severe and dry, there is often salt in the soil, and the Chang Tang is literally a desert. There are no woody plants, and very few flowering plants of any sort.

(ii) The outer plateau. Here the water is beginning to flow outwards, till eventually it reaches the sea;* but the watershed between the internal and external drainage is hardly perceptible, and it is not till the streams have grown larger that any change in the vegetation can be observed. Gradually the country becomes more fertile, with a steady slope to the south-east. Most of the Tibetan towns, such as Lhasa and Gyantse, are situated on the outer plateau. By the time they are reached the rivers are flowing in wide shallow troughs, and there are a few freshwater lakes. The climate is much milder than that of the Chang Tang, and there is a varied alpine flora; but there is no forest, and trees are confined to the water-courses.

(iii) The river gorge country. The streams which flow outwards from the rim of the Chang Tang have scooped shallow valleys for themselves. In order to escape from the plateau altogether however, and reach the plains below, they have to cut their way through the enormously high containing wall. This they do, giving rise to the river gorge country. Here we must altogether revise our ideas of Tibet. The altitude changes greatly within a few miles, so that the climate varies from warm temperate to cold temperate; but the atmosphere is always moister than on the plateau, and the precipitation greater. The river gorge country is divisible into two types: the upper gorge country, inside the main containing wall, with a comparatively dry, cold climate, a rich alpine flora and no forest; and the lower gorge country, outside the containing wall, with a much moister climate

---

* The rivers which flow outwards from the northern rim of the plateau flow into the Central Asian depression known as the Tarim basin.

INTRODUCTION

and densely forested mountains. It is important to remember that all three divisions are parts of the one plateau in different stages of disintegration.

With this brief introduction to the physiography of Tibet, I can pass on to the problem of the mountain ranges.

The eastern end of the great Himalayan range trends roughly east-north-east, and suddenly culminates in the high peak, Namcha Barwa (25,445 ft.), which overlooks the gorge of the Tsangpo. Here apparently it ends. Beyond this point the mountains trend, or appear to trend, north and south rather than east and west. However, a loop of lower mountains, built of the same materials as the Himalaya, and uplifted during the same period of mountain building, is found to curve abruptly south-westwards from here, and, continuing down the west coast of Burma, turn eastwards again through Java. Is this loop, known as the Malay Arc, to be regarded as the continuation, and the only continuation, of the Himalayan range? If not, can the Himalaya be traced eastwards beyond Namcha Barwa? Orthodox geographers say that the Malay Arc is the true continuation of the Himalaya, and deny that there is another. From time to time heretics have arisen who say that the Himalayan range is continued eastwards or south-eastwards across the mainland to the Pacific coast.

If that is so, what then becomes of the ranges which run from north to south between the Tsangpo river and the Yangtse Kiang? These mountains belong, so far as is known, to the much older Altaid system, which was in existence long before the Himalaya were uplifted. The answer is that the mountains in existence to-day are not primary ranges at all, but southward projecting spurs from a range which has been cut through by several rivers. The main axis of the Himalaya can still be traced through the nuclei of high peaks on these spurs or secondary ranges;* its general direction is then seen to run south-eastwards from Namcha Barwa.

As a result of many years' exploration in this corner of Asia, I believed that there existed a great range of snow peaks between the Tsangpo and the Salween, running in the direction it was assumed the Himalayan extension would follow. And so it proved. From

---

* They might also be called consequent ranges, since they depend upon the rivers having cut deep gorges across the main ranges.

the southern flank of this range springs the western branch of the Lohit river. Crossing this range was like crossing the Himalaya in Sikkim: one arrived on the plateau, with the climatic change which that involves. Much of this range was composed of sedimentary rocks, probably of the same age as those which outcrop on the Indo-Malayan mountains. They had been thrown into great folds by the Himalayan earth movements, and might therefore be considered as forming an eastern extension of the Himalayan range. Apart from this direct testimony, there is indirect evidence for the eastward extension of the Himalayan range, such as the presence of a great gorge on the Salween, south of the point where I reached it in 1933; the absence of any outstanding peaks except on the south-eastward alignment from Namcha Barwa; and the preponderance of an east-and-west over a north-and-south relationship in the mountain flora, in spite of the tremendous advantage the latter obtained in very recent geological time. For while the east-and-west continuity was interrupted by the presence of an ice cap, the mountain flora was being driven southwards by the same advancing ice, to mingle with the flora of the Indo-Malayan mountains. Yet when the ice retreated, the southern advance of the flora had made very little impression, while the continuity of the flora east and west was hardly impaired. This could only be due to the presence of great mountain ranges trending east and west, allowing the flora to spread in either direction.

I must now say something about glacial conditions in south-eastern Asia. Every traveller in the Himalaya has remarked on the retreating glaciers. But not every traveller has noticed that, between the Tsangpo and the Yangtse, the glaciation was so intense that a complete ice belt, almost an ice cap, extended along the marginal ranges of the plateau. In 1933 I found that Zayul had been completely glaciated to a point south of Rima. At the sources of the Irrawaddy, to the south-east of Rima, the ice came as far south as the 28th parallel. Farther east, on the Burma-Yunnan frontier, it came down to the 27th parallel. If the southern limit of the ice belt is plotted, it is found to extend in a direction roughly south-east from Namcha Barwa to the Likiang peak in north-western Yunnan.

Recent botanical exploration has shown that the affinity of the eastern Himalayan flora lies mainly with western China, across the river gorge country, and not with the Indo-Malayan mountain system. This fact is most simply explained by the existence of a former Himalayan range stretching eastwards into China.

In the course of my journey from the plains of Assam to the plateau of Tibet, I passed through many different belts of vegetation. In the Lohit valley dense evergreen tropical jungle covered the lower slopes of the hills, gradually becoming sub-tropical. In the neighbourhood of Rima this was replaced by Pine forest with *Iris Wattii, Lilium Wardii* and *Stellera chamaejasme* amongst the undergrowth. North of Rima, above the zone of Pine forest, was a belt of temperate forest, characterized by a mixture of Conifers (*Pinus excelsa, Picea Morinda, Tsuga yunnanensis, Abies Pindrow*) and broad-leafed trees, both deciduous (*Magnolia Campbellii*, Acers) and evergreen (*Rhododendron sinogrande*, Ilex). At still higher levels came a sub-arctic forest belt, including *Abies Delavayi, Larix Griffithii* and Birch, with an undergrowth of Rhododendron. Above the tree line was a scrub belt, consisting mainly of Rhododendron, but including also Acer, Sorbus, Syringa (Lilac) and Birch, the last alone of deciduous shrubs forming pure thickets. Above that again came the dwarf Rhododendron belt, consisting of a mixture of Rhododendrons, about ten species, with a few other shrubs, notably *Potentilla fruticosa, Rosa sericea*, Juniper, Lonicera and Prunus. Finally the alpine belt, with rock and meadow plants in great variety, such as Nomocharis, Primula, Gentiana, Meconopsis Iris, great numbers of Compositae (Aster, Saussurea), Ranunculaceae (Anemone *Paraquilegia microphylla*), Cruciferae, and many more. All these vegetation belts occur to the south of the great snowy range, and it is important to note that by the time we have passed from evergreen jungle to temperate forest, we have really passed from one floral region to another; from the Indo-Malayan region, as botanists call it, to the Eastern Asiatic region.

Crossing the great divide, another change takes place, and we find ourselves in the Central Asiatic region. No forest here—and the only tree *Picea lichiangensis*. Alpine flowers in variety and abundance cover the plateau; aromatic Labiatae, Compositae, Ranunculaceae, Saxifraga, Incarvillea, Gentiana, Pedicularis, Polygonum and many more may be mentioned; typical Central Asiatic genera are Astragalus, Oxytropis and Allium.

Included within the plateau flora is another occupying the hot dry valley of the Salween, to which may be applied the term arid flora. Shrubs are not entirely lacking, but they are stunted and thorny. The entire arid

INTRODUCTION

flora, including flowerless plants, comprises only about two hundred species. The shrubby plants include species of Caryopteris, Wikstroemia, Ceratostigma, *Sophora viciifolia*, Mentha and Aster. Amongst flowerless plants, the rosettes of *Selaginella involvens*, which curls itself up into a ball when the drought is prolonged, occur in thousands, and the fern *Cheilanthes farinosa* is often abundant. Herbaceous flowering plants are *Amphicome arguta, Didissandra lanuginosa, Onosma Hookeri,* Sedum.

Trees occur only where there is cultivation and irrigation; species of Salix, Crataegus, Juniper and a few others. This arid flora must not be confused with the flora of the outer plateau, which resembles it in appearance, but consists of very different species. In the deep river gorges, even at an altitude of 9000-11,000 ft., it is never very cold. The summer heat is intense, and the atmosphere dry. As on the plateau, a desiccating wind blows permanently. The soil is hard and stony. At times rain falls violently, but not for long; periods of drizzle also occur. The annual rainfall however is probably under ten inches. Southwards the arid, tree-less Salween valley passes abruptly into a forested region within a few miles, through a magnificent gorge. The inference is that the gorge has been cut through a great mountain range, and that the arid valley farther north owes its existence to the presence of this range.

It is also necessary to distinguish between the plateau flora, occupying a region so lofty and dry that all plants are necessarily alpines, and the alpine flora found on the high peaks of the forested river gorge country. The plateau flora is more closely related to the Central Asian floral region; the alpine flora of the Mishmi hills is Himalayan, Indo-Malayan, Chinese and in part indigenous. It may be called Sino-Himalayan. Typical plants, none of which are found on the plateau, are *Meconopsis betonicifolia, Primula Agleniana, Omphalogramma Forrestii, Cremanthodium Thomsoni* and *Nomocharis Souliei*. Most of the horticultural plants, of which I sent seeds to this country, were found on the plateau; that is to say they are alpines. A few came from the alpine ranges south of the plateau. The trees and shrubs came mostly from the temperate forest belt, Conifer forest belt, or scrub belt, of the river gorge country, south of the great divide; one only came from the arid region.

With this outline of the geographical problem, which I went to Tibet to investigate, and these few observations on the flora, I will pass on to the narrative of my adventures and discoveries.

CHAPTER I

UNADMINISTERED TERRITORY: JUNGLE WAYS

THE purpose of my recent journey to Tibet was, as formerly, to hunt for plants; to discover beautiful flowers and shrubs capable of being grown out of doors in this country and likely to thrive.[1] It was for this reason that my expedition was largely financed by the Royal Horticultural Society and a number of private horticulturists, each of whom pooled a sum of money in return for seeds I should bring back.

But while this was my primary object I intended also to collect plants as botanical specimens and to discover as much about the geography of these regions as I could, since most of the country through which I intended to pass was unexplored. The route I followed from Rima to Shugden Gompa had been travelled and briefly reported on by A. K.,* the pandit who in 1882, at a time when no European was allowed to enter Tibet, had spent three years in that country. Shugden itself, where I stayed three months, had been visited for one night by Capt. (now Lt.-Col.) F. M. Bailey in 1911. Bailey had followed a different route, and neither of these explorers had told us anything about the flora of Zayul. My journey to the Salween and most of my return route was through even less known country.

I had also a desire to investigate further a theory which my former journeys had led me to formulate, with regard to the orography of the river gorge country. Though my latest journey has not finally proved the correctness of my view, it has adduced to my mind very strong evidence that the current geographical theory must be rigorously modified. But this aspect of my exploration is expounded at length in the introduction and in the later chapters of the book.

I took with me two companions, Ronald Kaulback, a young Cambridge graduate, and Brooks-Carrington, the cameraman of

---

* Rai Bahadur Kishen Singh Milamwal, of the Survey of India.

Raycol British Corporation, who was to make a film of the country through which we passed. These two were permitted to accompany me only as far as Zayul, which belongs politically though not geographically to Tibet.

We travelled to India on separate ships and met in Calcutta towards the end of February 1933. From there we proceeded to Saikhoa Ghat by rail. Saikhoa Ghat is important merely as the railhead; it is a small village on the left bank of the Lohit; and to reach our destination Sadiya, eight miles distant, we had to cross the river. From Sadiya three great valleys lead to Tibet. The first of these, the valley of the Dihang, or Tsangpo, is impassable because of the hostility of the Abor tribes. The second along the Dibang valley is impracticable for lack of transport, since the valley is almost uninhabited and coolies unobtainable. The route we intended to take was up the Lohit valley, the eastern prong of the trident of rivers that go to form the Brahmaputra.

Two Mishmi clans[2] inhabit the Lohit valley, the Digaru and the Miju; between them there is a certain amount of friction. They are not now so savage and independent as the Abor clans; and there is a working agreement between them and the Indian Government, whereby they remain independent and pay no taxes, and in return for a small subsidy, the right to possess firearms, and the gift of a red sort of bath gown to each headman, they guarantee the safety of travellers in their country and keep open the jungle track to Rima.

In the cold weather, numbers of Mishmis descend to Sadiya, bringing goods to trade and looking for work, cutting cane for Indian contractors. In spring, they return to their own country and at this time of the year it is not difficult to get hold of coolies.

The fourth Mount Everest Expedition was assembling at Darjeeling. The news spread, and men flocked there from all parts hoping to be chosen for the great climb. The expedition had the pick of the men; but the local secretary of the Himalayan Club kindly undertook to choose three men for me who, though just too old to make the harder journey, were quite suitable for my purpose. They were all Bhutias, Buddhists by religion.

Tsumbi Sirdar was headman, responsible to me for the behaviour of the others. The Mishmis and Tibetans called him 'babu'. It is odd that to call a babu 'babu' has become almost an insult in Bengal; but to call a servant 'babu' is a compliment. When Mishmi children called *me* 'babu' I was furious, even though I knew their intention was good! Tsumbi was thirty-five and spoke Hindustani and Tibetan equally well. He had had experience with earlier Mount Everest expeditions but was now over age. He had a pleasant face and good presence; was cheerful and got on well with minor Tibetan priests and officials. But he was as bullying towards the Tibetan coolies as he was afraid of the Mishmis, when he discovered my hold over them was slight. Now and again he got drunk, and grew first abusive, then violent and finally unmanageable. But mostly he was sober.

Pinzo, an ex-soldier, was cook. He was short and sturdy. He had a raucous voice, which grumbled always. He had only to think of a complaint, to make it. His Hindustani was bad and his cooking worse. Altogether, he didn't seem to pull his weight.

But he did pull it. He did more work than any of the others. He was not afraid of hard work, but he was annoyed at it. And he could stay the course. He had been a *naik* in the Indian army, on field service in Persia, and spoke a little English which he had picked up from his British officers. He was the best all-round man I had, reliable and intelligent; so that when I had to choose the safest man to take Ronald and Brooks-Carrington back to India, it was Pinzo who had to go, much to my regret.

The third man was Tashi Tendu, a stocky fellow with a mop of black hair. Though he looked like a Belcher charwoman, he had been one of the 'Everest Tigers', and had carried loads to over 25,000 feet both on Mount Everest and on Kinchinjunga. Now that he was too old for such strenuous work, he willingly joined an expedition to Tibet as man-of-all-work. He didn't like a quiet life.

Cause and effect had no meaning for him; two and two, if he ever put them together, always made five. He spoke little Hindustani and less Tibetan; altogether he was rather what Americans call 'dumb'. But he was the best-tempered and most willing worker I've ever employed. There was nothing he wouldn't turn his hand to, and very little he didn't have to turn his hand to. That is the advantage of Buddhist servants; they have no 'caste' or religious fads. Only once in Tibet did my servants ask for a special holiday; and that was to go to burn incense at a famous shrine near Shugden Gompa.

All three of them left wives in Darjeeling, and arrangements had to be made to pay monthly family allotments from their wages. While they were in Tibet not once did they communicate with their wives. But then my own wife went four months without news of me, and I myself was cut off from the greater world for seven months. It proved easier to send letters out of Tibet than to get them in. The Lhasa Government was kind but not helpful. There is a postal service from India to Lhasa, via Gangtok and Gyantse. I suggested that letters should be sent to Lhasa and thence forwarded via Chamdo to me at Shugden Gompa, 400 miles away. When the Indian Government requested the Tibetan authorities to do this, they replied that they would be pleased to do so, but that they didn't quite know where Shugden Gompa was!

We spent several days in Sadiya, dividing the baggage into 60 lb. loads, buying rations, and arranging for transport, with the help of the Political Officer, without whom we could have done nothing. On March 9th the Political Officer drove us in his car as far as Denning (45 miles), where the cart road ends. The bulk of our luggage had been sent ahead by bullock cart; and the village bus followed behind us with our personal baggage and servants.

The coolies for the first stage of our journey were recruited from the Digaru Mishmis. They were sixty-five in number (not counting the advance party), and were nominally in the charge of an aged *gam* or headman called Nimnoo. But this did not mean much: because the Mishmis are very independent people and the most influential *gam* has little control over his flock.

We spent the night at Denning and set out early the next morning. It was raining. It is almost always raining in the Mishmi hills, especially during the summer.[3] The soil is being continuously washed down into the valleys. A landslide may dam the torrent and form a lake until the pent-up force of the water breaks the dam and floods spread destruction on the plain below.

From Denning we began to climb by a zigzag, switchback path upon the mountain side. Thick jungle lines the path. Primary jungle is fairly easy to cope with, but when the jungle is cleared by burning and cultivation, and then abandoned, in a few years impenetrable thickets spring up much denser than before. This secondary growth is so wet and green it cannot be burnt. After about ten years it has grown sufficiently to be cut and burnt again.

To travel these hills at scenic-railway pace would be exhilarating, so abruptly the path twists, dips, climbs and changes direction. But the pace of marching is very tiring; and the path is so enclosed by jungle that seldom can one get glimpses of the mudflat plain below.

The path, which is called the Lohit Valley Road,[4] mounts 2850 feet to Dreyi, an aerie on a spur—where we spent two nights owing to coolie difficulties—and then up again to the Tidding Saddle (6000 ft.), zig-zagging all the time.

From there it winds down 4000 feet to Theronliang on the Tidding River, where stands the last bungalow in India at mile 72. From there to Rima is 200 miles by the river through unadministered territory. The path is just a rough mountain track, along which pack animals cannot travel. All transport is human. What is good enough for goats and Mishmis has to content botanists and explorers.

At Theronliang I had a long consultation with the *gams*, during which the coolies, men and women alike, hung round smoking and interrupting when they chose. The discussion, which seemed interminable, ended in an agreement that Nimnoo would accompany us fifty miles up the valley and then hand us over to an influential *gam*, called Jaglum, a Miju Mishmi. Jaglum would take us on to Rima.

After having sent a last communication to Sadiya, by telephone, we set out early next morning into unadministered territory. It was a fine day and in the afternoon we reached the gorge of the Lohit. When we made our first camp at dusk, Ronald urged that we should sleep in the open as it was such a fine evening. I should have known better, having had experience of the Mishmi hills. However, I agreed. When we retired the stars were bright overhead; but at midnight there was a rumble of thunder, and before we were properly awake we were drenched. After that sleep was intermittent, as puddles gathered round us. I was heartily glad when day came: I do not mind what discomfort of cold, wet or hunger I endure during the day so long as I am sure of a warm, dry bed. But the others took it cheerfully. It rained pitilessly all the next day until we reached our camping ground. Then it providentially cleared, the sun shone out and we were able to dry our bedding.

By doing this part of our journey in early spring we avoided the worst discomforts of jungle travel. On wet days we were slightly troubled by leeches; and there were of course the usual biting flies, the sand fly and the blister fly. But none of these was anything like as bad as they would have proved if we had been travelling even a month later.

On the sixth day after leaving Theronliang we arrived at Pangum, where we were greeted in a loud voice by Jaglum. He treated us as if we were lepers, who must at all costs be isolated from the village. I told him we would choose our own camping ground; and, taking his hint, chose to camp at some distance from the village. There is plenty of room in the Lohit valley.

Next morning Nimnoo came to say good-bye. He had done his bit; and though he had grumbled and cadged, he gave no trouble and produced coolies as required. There are few villages in the Lohit valley, and they rarely contain more than three or four huts apiece, so that in supplying coolies without a hitch he had done well. I was grateful to him, despite his fawning for cigarettes, sugar and pinches of tea. He raised his arm to me in a sort of Nazi salute and waddled off, a small boy trailing behind with his blunderbuss. On my return I heard that he was dead.

Miju and Digaru were jealous of one another: between all Mishmi tribes there is as much squabbling as between Scottish clans of old. The Digaru are pro-British, migrating to India for the winter. The Mijus are pro-Tibetan and make for Rima when they want work.

I got on well with Nimnoo after I learnt to treat him like a naughty child. Jaglum was different: a short, powerful man with a grim face and a swagger. The only time I ever saw him smile was when Brooks-Carrington filmed him in Rima. He tried to wheedle me by saying that while the *gams* were content to obey the *sircar* (Indian Government) the young hotheads were not. He had to be tactful, and more than hinted that he could be more tactful if I gave him cigarettes and tea.

He had the oriental aversion from responsibility, but when things went awry, no occidental skill in blaming the system. He had no need to do so; there were his coolies he could blame. He himself would be delighted to do what I wanted!

When I asked how far it was to Rima, he said eight days' march. I was paying my coolies by the day—at twice the normal rate. Two days later it was still eight days; and when I demanded an explanation he said, '*My* coolies could do it in eight days. But the *gam* of Halei is now responsible, and *his* coolies say Rima is eight marches from their village.'

That was the trouble. No coolie would do more than a day's march. They wanted to return to their fields. They made the march as short as possible, collared two meals of rice and their pay and set off home.

UNADMINISTERED TERRITORY

For a week I gave way to Jaglum in everything. Then I called his bluff. I had insisted on the coolies doing four days' march in four days. Having dawdled for three days they had to march fifteen miles on the fourth. It was later than usual when we camped. The advance party, having no one to drive them on, had made five days of it and eaten two bags of rice over and above their allowance. Jaglum demanded five days' pay for them, and five for the coolies with us, to equalize the position. I equalized it by giving him four days' pay for each. Seeing I was adamant Jaglum said wouldn't I give the poor coolies a present, eight annas each? (A shake of the head.) Six annas, then? (Another shake.) A beggarly four annas? But I paid him nothing: and he went away sulking; though he cheered up when I paid Ahu Lomei, the one-eyed *gam* of Minzong, for a hut he built and the present of a tough cock. He tried to get me to pay for repairs to the road and the bridge, but I pointed out that that was implicit in the system of 'loose political control'.

When Ahu Lomei, who was hung with chains of rupees like a city alderman, asked for a commission on all coolies he supplied, I refused that impost too. I knew the coolies were only too willing to carry at the price, and if he forbade them to carry they would disobey him. The *gam* threatened me, saying that if he got no present, I would get no coolies. I told him I never listened to threats and sent him away.

Tsumbi, who was scared by Mishmi swashbuckling, swore the coolies wouldn't dare to come. I told him the *gam* wouldn't dare keep them away; and so it proved. They all turned up the next day, and I took exquisite pleasure in giving the *gam* a present; at which he was so delighted he followed us all the way to Rima.

Meanwhile the advance coolies quarrelled with my coolies for not having taken five days to do four marches, thereby losing everyone a day's pay. I thought they were coming to blows. But the storm blew over.

At Rima I fined Lomei's coolies for stealing a water proof sheet. He was so distressed that he went off and got drunk and stayed drunk three days. However, we parted good friends.

The advance coolies had been instructed to drop six loads for us at Minzong, and these loads were carefully marked to avoid mistake. Six loads were handed over to us by the headman, but of them only two were right. I also discovered that a case of oil I had ordered in Sadiya, meaning kerosene, was sweet cooking oil, good enough for tough chicken or lubricating a rope bridge, but not much use for heating and lighting.

We had wet and fine days during the march. When it was very wet Jaglum came to my tent in the morning and said that the coolies could not march as the cliffs were too slippery. If I anticipated him and said the weather was so bad we would rather not march, he answered truculently, 'If the coolies don't march, they'll return to their villages and not come back.'

There was only one way to combat this obstinacy. If we decided to rest a day, I made feverish preparations for departure; if we wanted to march, I lay abed, thereby allowing myself to be bullied into acquiescing in my real wishes.

One night the Mishmi coolies were caught in the open as we had been. Usually they built huts of bamboo, roofed with banana leaves and lined with grass. A thunderstorm broke overhead in the middle of the night; and in a moment the camp was filled with cries as the coolies, swarming like angry bees and waving pine torches, ran helter-skelter for cover.

A great deal of opium poppy is cultivated in the Lohit valley. We passed through fields of it every day. Bootleggers buy the opium cheaply and smuggle it into Assam to supply the insistent demand there due to restriction. Some of them get caught; but the trade is remunerative and prosperous.

Not all opium grown is destined for the Indian market. Much is used locally as a prophylactic against fever. It is not smoked as opium: chopped tobacco is steeped in an aqueous concoction of opium and smoked in the ordinary way. Moreover opium is a pain-killer and is often used by the Mishmis to stave off hunger or fortify the body against collapse under the strain of exceptional fatigue.

Every year more land is being put under opium cultivation and less devoted to food crops, with the result that the Mishmis are coming to depend more and more for the provision of their food supply on the export of opium. But it is not unlikely that the poppy can be grown on land temporarily unfit for anything else, and that it contributes to an improved rotation, thereby doubling the area of arable land.

Though the Mishmis are becoming tea-drinkers, or at least have the desire for tea, they make no attempt to cultivate the tea plant, which would satisfy their own needs and provide a handsome trade with the people of Zayul. The tea plant grows well in the hill jungles of Northern Burma, where the Kachins pickle the leaves in bamboo tubes. There is no reason why the Mishmis, enjoying a similar climate, should not do the same.

When I reached Rima, however, I found a ready-made local tea industry. That tea would grow in the Mishmi hills is likely; that it should grow in Rima was incredible. They told me that the industry had been started in response to urgent representation from the Government—no doubt there had been a hitch in the brick tea trade with China. A Tibetan needs tea as an Englishman needs bacon and eggs; so an official said, 'Let there be tea', and there *was* tea. It was quite biblical.

Having met a formidable caravan of 600 yak carrying tea to Chamdo, I investigated the tea industry on my return to Rima in December. The assumption underlying the business was logically unsound; the old fallacy of A is B, therefore B is A. China tea is in bricks; therefore what is in bricks is tea. Once there were three grades; now there are two. Perhaps the tea grades will end like the ten little nigger boys. Third quality tea was withdrawn from the market because people would not buy it, as it made them ill. It was made from the leaves of the Indian Alder (*Alnus nepalensis*), which is scarcely distinguishable from our alder, and will grow anywhere where the soil is poor enough. It is one of the first trees to establish itself after a fire or landslide, or when land falls out of cultivation.

The first grade is made from *Loranthus odoratus,* a relative of our Mistletoe; a second from a shrub called *Osyris arborea* (Santalaceae), they are made into 'tea' simply by the leaves and twigs being compressed into bricks.

As the Tibetans emulsify their tea with rancid butter and then neutralize the flavour of butyric acid with salt, there is not much difference in taste, whatever 'tea' is used. The chemical constituents of tea and 'tea', however, are very different. In fact analysis shows that both Loranthus and Osyris contain only a trace of tannin such as is found in every plant, and it remains to be seen what will be the reactions of infusing an oleaginous salt solution with extracts of these leaves.

At Minzong we entered the territory of the Miju Mishmis, who are similar physically to the Digarus, though they speak a distinct dialect.

Tibetan influence is spreading among the Mijus. The cane rope bridge, over which the Mishmi painfully drags his body suspended in a ring of twisted cane, has given way to the bamboo rope, with a heavily greased wooden slider. Prayer flags flutter in the villages and the Tibetan monks will soon venture among the people, turning them from primitive animism to Buddhism.

It was five marches from Minzong to the frontier of Zayul, and another three to Rima. Bad weather continued to delay us; but a friendly *gam* had constructed such a comfortable camp for us two marches from Minzong, with grass huts in a clearing, that I did not mind having to spend a day there.

CHAPTER II

THE HOT VALLEY

On March 29th, we suddenly emerged from the forest on to an open grass-covered terrace whence we first saw the snow-clad mountains of Zayul ahead. That afternoon we crossed the river by rope bridge. The Lohit is here sixty yards wide and the rope sagged badly in the middle. But the natives tightened it till it had a definite slope downwards from our bank. We lubricated it thoroughly with our excess cooking oil and shot across the river at high speed, landing on the other side with a bump. That night we camped in the first Tibetan village.

The landscape changed.[1] The jungle was left behind. The valley floor was bare and rocky, though no more than 4000 ft. above sea-level, and between the river and the precipitous mountains on either side the ground rose in narrow terraces, which were covered with Pine forest. A little above the rope bridge, an immense moraine blocks the valley, showing that once a great glacier descended at least as low as this. To-day you must climb thousands of feet to find even the remnant of a glacier so far to the south.

We halted a day's march from Rima and I sent Tsumbi ahead to announce our arrival. The Governor, who was then in residence at the 'winter palace', sent five ponies to the edge of the plain for ourselves and our servants. Thus we rode the last four miles to Rima in gratitude and comfort. It was April 1st.

Our arrival at Rima[2] was something of a triumph. The whole population, men, women and children, numbering about 200 persons, lined the path up which we rode to the headman's house. But they did not cheer or, in fact, display any emotion whatsoever: so perhaps it was not much of a triumph after all. Servants, cap in hand, held our ponies while we dismounted, and the headman, a tall burly figure with a straggly beard and cropped iron-grey hair, led us into his house. We sat cross-legged on a divan before a low wooden table, on which were placed china bowls, a dish of walnuts, and another of dried persimmons. Buttered tea was prepared, and we sipped it dutifully. Then our host poured us out some fiery rice spirit. The headman's servant, a small sharp-featured little man, who had a habit of putting his head on one side like a contemplative hen when he looked at you, had been in Darjeeling. He spoke a little Hindustani, and even a few words of Chinese. We met the headman's wife, a plump, pleasant-looking woman, shapelessly dressed in a Tibetan shirt-waist, with striped apron and fur-lined cap. She wore large turquoise and silver earrings, which contrasted pleasantly with her rosy cheeks. Meanwhile quarters were being prepared for us. The headman's house, the travellers' rest house, and the official residence stood close together, enclosing an open space where Ronald and Brooks-Carrington pitched their tents; while I camped in one room of the rest house—the other being taken over by the servants, and used also as our kitchen. After a night's rain, in the course of which I got a wetting, I vacated the guest house. An even more dilapidated shed close by then became my bedroom, giving us more room in the communal apartment where we took our meals.

When we were shown to our quarters, I told Tsumbi to prepare for visitors. Then I sent my credentials and presents to the Governor, accompanied by a ceremonial white scarf, thanking him for the ponies, and asking when it would be convenient for me to call. Amongst the presents was a small alarm clock, an electric hand-torch, a box of scented soap and a pocket-knife. The Governor returned a courteous message to the effect that he would pay me a visit; shortly afterwards Tsumbi whispered that he was coming, and I had tea made, English fashion, with condensed milk and sugar. The Governor of Zayul[3] was then announced. He arrived with his servants, bringing presents—a bag of rice, a bag of flour, and a bag of walnuts. He was a pleasant-looking man of about thirty-five, rather bulky, smooth-faced, with an easy manner. He spoke rapidly in a low voice, emphasizing his words with gestures. With him came his assistant, who could not have been more than twenty years of age, and looked like a schoolboy. Evidently he was not used to society, for he was shy and awkward. However, the Governor seemed quite at home—he was a Lhasa man, of good family—drank his tea, with milk and sugar, and helped himself to biscuits. The visit was a distinct success.

I took a liking to the Governor, and we became good friends. But he was surprised to see us; evidently the Lhasa Government had not notified him that we were coming. He was astonished when I told him that we had walked from Assam. It seemed to him an insane thing to do, because the wealthy and official classes never walk; they ride. 'Only beggars walk in Tibet', is a well-known saying in Kam, where the pack-roads are exceptionally good—for pack-roads. He asked what we were doing, where we wished to go, and how long we intended to stay; and when I told him my plans, he promised to assist me. No sooner had the Governor departed than Tashi came to me with a broad smile on his weather-beaten face, and spoke confidentially: 'I was his servant in Lhasa for a year, *sahib*. He knows me. He is very friendly.'

I thought it best to stay here some days in order to consolidate my friendship with the Governor. Also here was a heaven-sent opportunity for Brooks-Carrington to get some shots of Tibetan life. But we could not leave at present in any case, because the loads sent from Denning in advance had not yet arrived. We had passed them on the other side of the river, abandoned by the Mishmis two days short of Rima and ignored by the Tibetans. It was not till I had asked the Governor to send back instructions for the loads to be brought in that they arrived ten days after us. So much for luggage in advance in the Mishmi hills.

At this season Rima was quite a busy place. Most of the headmen from the up-river villages had come in, bringing tribute in kind, and there was also a sort of daily market, though one could buy little except rice and walnuts. There are no shops in Rima.

With two of the up-river headmen we made friends. They came from villages we frequently heard mentioned, and we were destined to meet them again. The first headman was Wunju, of Giwang, a village of good repute, and a likely place for us to stay, to collect plants. Wunju was a little man with sharp, eager features, a husky, plaintive voice, and an exuberantly polite manner. He wore a sugar-loaf Punchinello-like hat, which he snatched off hurriedly whenever he met us, at the same time bowing low and protruding his tongue in the approved Tibetan manner; not, like some people, sticking it out when our backs were turned. Wunju invited us to his house at Giwang. He said it was a fine village with lots of rice terraces, and that we should be very comfortable there. I asked him if there were any flowers. That puzzled him, and he admitted he hadn't seen any, but added tactfully that if I went there doubtless I should find some.

The second headman was Kyipu of Modung. He was a large sleek man, almost corpulent, with a fat smiling oily face. Tsumbi called him a successful box-walla*, and indeed he looked it; he did himself well, and had great possessions. But he welcomed us, invited us to Modung, and gave me a good deal of information about the Rong Tö valley.

Although Rima had been visited by three or four Europeans before us, no European had ever been up the western branch of the Lohit river, called the Rong Tö Chu. For information I relied entirely on the report given by the *pandit* Kishen Singh—excellent, so far as it went—which I was able to check. It was rather surprising how little anything had changed in fifty years. One expected to find villages for the most part still where they stood half a century ago, but I was able to corroborate almost every detail of the narrative; and Kishen Singh was an observant man.

Though there are no regular shops in Rima, many families had a few things for sale—milk and butter, for example, occasionally eggs, or a lean fowl, or a piece of pork. Chinese sugar, too, could be bought—at a price. There was a clandestine traffic in the vilest cigarettes from India; clandestine because the Governor, very wisely no doubt, had forbidden the smoking of these noxious weeds. In a quiet way indeed the Governor was something of a reformer, and zealous too. He it was who instituted the curfew in Rima, in the form of a man who went round the village every night after dark, warning people to put out their lights and go to bed. This was a precaution against fire. He was anxious to encourage trade with Sadiya; and more than one headman asked me to curb the rapacity of the Mishmis, whom they found troublesome to peaceful traders. I was to make the world safe for democracy. But realizing that all efforts in that direction elsewhere had only succeeded in making democracy safe for the next world, I gently declined. I need hardly add that it was none of my business; we had enough troubles of our own with the Mishmis. Nor is the fault entirely on one side.

The authority of the Tibetan Government in lower Zayul is shadowy. To begin with, there are comparatively few Tibetans in the country. The indigenous inhabitants are a pygmy jungle tribe who, during the last hundred and fifty years, have become Tibetanized by colonists from

---

* Box-walla: a rather derisive term in India for a business man, as opposed to a Government servant.

the north. It is the lamas who are chiefly responsible for the civilization of the aborigines; and the land-owning class who are responsible for a measure of absorption. But there is no mistaking the aboriginal pygmy strain. The women are particularly coarse-featured and in figure amorphous. The Governor told me he had the greatest difficulty, when he first came, in understanding the speech of the people. He had only lately been appointed to Zayul and was not pleased; but I have no doubt he was sent there from Lhasa at a time of strained relations with the adjacent Chinese province of Szechuan, because he was considered an able man. His predecessor had died suddenly, and his widow still occupied a room in Government House.

There is no post office at Rima. The Tibetan mail service does not extend beyond Chamdo in this direction. The Governor, however, arranged to send a man to Sadiya with our letters, and at the same time he wrote a friendly letter to the Political Officer, thereby establishing official contact. On hearing the good news we wrote letters feverishly; these in due course reached Sadiya, and thence England.

Government dispatches between Chamdo and Rima are sent by mounted courier. A dispatch rider arrived during our stay. He was a sturdy, hard-bitten man, armed with a Lee-Enfield rifle of the year 1903 and a large bandolier. He rode a shaggy pony, his rations and bedding following on another pony, in charge of a servant. Rifles are more easily acquired in Tibet than ammunition. Nevertheless, the day after the courier arrived the Governor brought out his rifle and a few cartridges, and the three of us indulged in a little target practice, to the delight of a crowd of local sightseers. But ammunition was far too precious to squander, and we fired no more than three rounds each. I noticed that the Governor forgot to clean his rifle.

On April 12th the Governor invited us to dinner. There were china bowls of chickens' livers, spinach, and pork, washed down with buttered tea and raw spirit drunk out of jade cups. Best of all were the dumplings of maize meal with a slab of honey on top, and boiling butter poured over them. The Governor, who ate sparingly, was in merry mood. Butter lamps flickered on the family altar, dimly illuminating the wall behind and a row of wooden pegs from which hung rifles, wooden saddles, bridles, leather whips and other appurtenances of rank. Our own red silk shaded candle lamps, which had been requisitioned for the feast, cast a warm glow over the table.

When our luggage arrived I got down to the question of departure, and the Governor promised to make arrangements for transport. We would go first to Giwang, four stages up the Rong Tö valley; then to Solé, another two stages. I changed some rupees into *tankas,* the only Tibetan coin current, at the rate of six *tankas* for a rupee; that made the *tanka* worth about tuppence-ha'penny. The date of departure was fixed first for the 16th, subsequently postponed till the 18th.

Meanwhile, the people of Rima were not inactive. A troupe of Kampa dancers arrived and enlivened the village. Tibetan dancing is dynamic, cymbals, drums and a primitive type of very squeaky fiddle giving rise to a medley rather than melody. There is not much rhythm, but the dancers work themselves up into a hysteria of gymnastic motion. Then came a religious troupe, who demonstrated how to cast out devils, even the most stubborn. The face of the possessed man was concealed behind a paper mask. He gyrated violently, till even I felt giddy watching him. The devil, as susceptible to weird music as a serpent, shook him to the foundations, but eventually emerged, leaving the late victim exhausted.

The Tibetans put duty to their religion before everything. Thanks to the zeal of the lamas, Buddhist ritual has spread throughout Zayul, and penetrated even into the Mishmi hills. Every householder in Rima daily offered the fragrance of imaginary burnt offerings to heaven, a rich volume of aromatic smoke, obtained by burning bundles of an abundant weed called Artemisia (Keating's Powder is made from it) in a small structure like a clay oven, ascending into the air. Every day men, and women too, marched solemnly round the square temple (which contained a large image of Buddha) muttering prayers and twirling prayer drums. As already remarked, the population of Zayul consists of two distinct elements, Tibetan and aboriginal. The aborigines must have swarmed off some common stock long ago, and remained in undisturbed possession of Zayul until comparatively late years. They are pygmies, and now form the serf population. The Tibetans are Kampas from eastern Tibet, a tall swarthy people, full of good cheer. Kampa women are remarkably handsome, their piercing black eyes and raven hair, together with their warm colouring, suggesting a gipsy strain. They love a gay life brightened by song and dance with plenty of drink. Much of the family wealth is invested in jewellery—large silver earrings set with coral and turquoise, finger-rings, and

A KAMPA WOMAN

silver bangles. A collar of blue beads is often worn, and a silver charm box suspended round the neck. In the deep warm valleys of Kam, the women can indulge their taste for bright blue cotton skirt and jacket; they are not, like the people of the plateau, compelled by the cold to dress in thick woollen *chupa*, or sheepskin. Very different are the serfs—shapeless pygmies with fat round face and wide nose. They dress in drab sacking, and for decoration wear only a multitude of bright yellow bead necklaces—the badge of the serf, it would seem. Also it hides the goitrous lump in the neck.

There is no forest below 6000 feet at Rima, the river banks being lined with a thick growth of shrubs, both evergreen and deciduous; a sort of transition between the hill jungle of the Lohit valley below and the temperate forest of the higher ranges. Here I found *Ceratostigma Griffithii, Prunus Padus* (the Bird Cherry), *Tupidanthus calyptratus, Pittosporum floribundum, Rosa bracteata, Rubus lasiocarpus,* and *R. moluccanus,* together with species of Philadelphus, Deutzia, Macaranga, Viburnum, Indigofera, and the climbers Holboellia and Zanthoxylum. In this comparatively dry climate many shrubs have fragrant flowers or aromatic foliage. Above the village the dry rocky slopes are covered with *Pinus Khasia* and evergreen scrub oak. Only in the deep sheltered ravines, above 7000 feet, do we begin to glimpse the temperate rain forest, rich in Laurels (Litsaea), Willows, Clematis, and the gorgeous Carmine Cherry. This last is one of the most lovely flowering trees in existence. Imagine a Cherry tree eighty feet high, not yet in leaf. The deep carmine buds, in clusters, droop over gracefully and burst like red stars showered from a rocket till the whole tree is enveloped in a deep carmine cloud. When the setting sun shines through the branches, lighting up the red flower clusters in an incandescent glow, the effect is superb. About May, when the flowers are over, the large leaves come out, and stay on the tree till October. I first saw the Carmine Cherry in far Northern Burma, some forty miles southeast of Rima, but on the other side of a great mountain range, in 1931, and collected seed of it. Very few of the seeds germinated and to-day there are only about a dozen young trees in England. At Rima I saw one tree; but there is no doubt the species is widely scattered over the high frontier ranges of Burma-Assam. An avenue of the Carmine Cherry would be a sight for the gods; should it prove hardy, as it appears to be in the south of England, a collector might well be commissioned to go East merely to collect seed of that one tree.

THE HOT VALLEY

Thus by the time Rima is reached, the flora of the Lohit valley has fundamentally changed from subtropical evergreen rain forest to Pine forest. The climate is both drier and more extreme than that of Sadiya, being hotter in summer and colder in winter, when snow actually falls, though it does not lie long in the valley. The dryness is due to a hot air draught drawn up through the Lohit gorge from the plains; but this draught is confined to the gorge, and temperate forest prevails a couple of thousand feet above the river.

Below Rima the gorge opens out and there are extensive gravel terraces on both banks. One such terrace is 1200 feet above Rima. Some of the high terraces were once cultivated, but owing to irrigation difficulties they have been abandoned.

Half a mile above the village the Lohit river divides into two branches. Up the eastern branch lies the main road to Chamdo and to China. Our object was to follow the western branch. Just below the confluence is a rope bridge, giving access to the western river or Rong Tö Chu.[4]

The night before we were due to start, the Governor gave a small dinner party in our honour, the other guests being the widow of the late Governor, the Governor's pretty Kampa wife, and the junior official. We enjoyed a rich meal of pork and sauerkraut, washed down with copious draughts of rice beer. As a farewell gesture I played the ukelele to our delighted host, while Ronald performed an extempore dance which would have gone down well enough at a Piccadilly night club, but which created a huge sensation in official circles in Rima. The Governor was not less delighted than was Herod with Salome; later he himself had occasion to offer me a head on a charger.

I left the Presence at 10.30, feeling slightly unwell, and retired to my room, where I gave Tsumbi instructions for the next day. Then I turned in. Two hours later I awoke and was violently ill. After an uncomfortable night, I was glad to postpone the start, owing to a shortage of coolies; for I was too unwell to travel. On the 18th I got up, still feeling more dead than alive. We crossed the river by the rope bridge, a long operation, and for the first time had a good view of the unknown valley. Exploration had begun.

CHAPTER III

THE VALLEY WHERE NO MAN GOES

A narrowing valley stretched northwards in front of us, and was cut off abruptly by the converging mountain ranges. But the river came thundering out of a deep gash, and a mist of spray hung in the air. Beyond, the cobalt sky shut down tight, like a lid. I felt rather awestruck as I gazed, noting the puffs of bright cloud shining against the violet hills. No white man had ever seen this valley before. The river was not less than a hundred miles long....

After the first few miles it looked as though there could be neither villages nor cultivation. The mountains straightened their backs and approached each other more closely, squeezing the river between them till it roared. They were covered with Pine forest below; but the deep shady glens and cool north slopes bore a richer covering of Oaks and Maples, Rhododendrons and Magnolias, some of which were in bloom. The tops of the hills were white with snow, and the air was cool and moist. Dawa Tsering, the Moon Man, had told me of two villages where he advised us to stay in order to collect plants and take photographs. The first was called Giwang, and it was four stages beyond Rima. The second, called Solé, was two stages further. We therefore made for Giwang; but it took us more than four days to get there, because we required so much transport that no village could supply it all at once. This, however, emphasized the sparseness and poverty of the population, rather than any sybaritism on our part.[1]

The woods were swamped in a lagoon of pale violet flag irises.[2] When the wind ruffled them, they danced like butterflies on a green baize cloth. Here the tall reeds of the winter flowering *Mahonia calamicaulis* looked like miniature Palm trees; they grew in clumps twelve feet high, the ringed stems leaning at all angles. A wide collar of monstrous leaves, stiff and sharp-pointed, surrounded a bunch of slender stalks, which sprang from the centre, and gradually lolled under the weight of grape-coloured berries. Men eat the berries. They are said to cure headache; but I found them astringent. The flowers, which are bright yellow and fragrant, spouting out from the centre of the crown in an immense fountain, open before Christmas, and the fruits ripen the following summer, during the rains. The young leaves, which appear about May, are polished salmon-pink and quite soft, though they have the same formidable spear points as the hardened weapons. Thus *Mahonia calamicaulis* has something to offer nearly all the year round, including its crinoid shape.

At the mouth of the valley there were paddy fields down by the river, and villages up on the bluff. But presently the valley contracted. Gravel terraces sloped from the foot of the mountains to the brink of the gorge, and were cut into wedges by streams which drained the western hills. These terraces were covered with Pine forest, in which men had made clearings and built log huts. Here they tilled the rough soil with a wooden plough drawn by two cattle, and sowed wheat and corn. Larger villages were scattered down the wide alluvial fans broadcast from the mountains by intermittent torrents. Men stepped the earth fans, patiently, and grew rice, irrigating it by means of channels and flumes. The path lay over the terrace blocks, through endless forests of Pine, plunging suddenly down 200 feet into a ravine, and up the other cliff; then turned aside to ascend an earth fan to a village on the bluff. The villages were small, but they were not far apart; and if we met no one on the road it was only because no stranger ever comes to this valley. The people were working in the fields.

Perhaps the Pine forest had not always been so flowerless and empty and fragrant as it was now. Before men came here and cultivated the soil and kept cattle and ponies, there might have been flowering shrubs and herbs. But now every winter they fire the Pine forest belt, so as to stimulate the new growth of bracken and grass when the rain comes, and provide fodder for the animals which eke out a miserable, saltless existence in the valley. This annual ignition prevents many plants which would otherwise grow here from getting a hold, and stereotypes the flora into a few fire-proof species.

Zayul is still in process of colonization by Tibetans. The aborigines were a race of pygmy forest-dwelling people, with moon faces. They have for the most part been Tibetanized, and now form the slave population, but being more under-nourished as slaves than they were as freemen, nothing has been added to their stature. Nevertheless they are garbed like Tibetans of the poorest class. They wear the national hempen dressing gown, called a *chupa*, to which in the cold weather is added a sleeveless jacket of red gooral or goatskin. The women have a curious passion for clouded yellow, and adorn their goitrous necks with ropes of opaque glass beads. Hewers of wood and drawers of water, they nevertheless eat their coarse food squatting round the rich man's fireside with the rich man's family in the dark and noisome kitchen. They work all day in the fields. Often I heard them pounding paddy in the yard at ten o'clock at night by the light of a spluttering pine torch. They carried our loads for us, queer human beasts of burden, shuffling along the stony path.

They never had any money. Even the people of the valley who were not slaves never had any money before we came. They were in perpetual debt—all except a few rich Tibetan chieftains. They borrowed money from the officials at Shigatang, to buy a plough, or a yoke of oxen, or seed to yield a crop, or a piece of land, and the rate of interest compares favourably with that charged in the highest financial circles: 120 per cent per annum. This business is largely controlled by Tibetan women of quality, widows of former officials, who have settled in Zayul, or wives of rich traders; it is so lucrative that every widow gladly lends her mite. The officials at Rima also take a hand, and supplement their meagre fortunes by usury. They lend money to the farmers in spring on the security of their crops. In the winter they ride round from village to village, from house to house, demanding the interest due. Of course it is never forthcoming—it was a bad season, the crop failed; or it was a good season, but the government assessed taxation so high that there was no surplus.... Anyhow, the farmer has no money. Then events follow a regular course: persuasion...intimidation...castigation...the farmer pays an instalment, and the debt is carried forward. It would however be a mistake to suppose that our advent enriched the small farmers. Whenever I paid out money, it was instantly claimed by someone else. As for the slaves, they had owners, who owned likewise their work. We merely attracted several rather unsavoury money lenders to the valley, who smelt our rupees as the South American *piranha* is said to smell blood. We tainted the whole valley with a stream of silver, and the *piranhas* followed in our wake. When I returned eight months later they had drained the valley.

Much of the Tibetan influence is derived from the slow but irresistible power of the church. Missionary work is zealously carried on amongst the slave population. Truly it keeps them poor; but the slaves are wise enough to know that anything they have to pay for is probably worth having. The something-for-nothing merchant would quickly starve in Tibet. Every village has its monastery, even if there is no incumbent; and a shrine every few miles along the road reminds the traveller that *O mani padme hum* is what he needs. It is the only poster he ever sees in Tibet—a poster in stone; the church has a monopoly of propaganda. But it would be a mistake either to ridicule or to underrate ecclesiastical influence. The church is Tibet. It is the one unifying, cohesive force which holds together the people of outer space. Tibet as a colonizing influence may seem to be a new idea, slightly laughable. But it is not, in fact, new. Tibet has been colonizing, eastwards and southwards, for centuries. First the traders, then the missionaries. And the new religion they bring is vastly superior to the old tribal religion; it substitutes horse sense for nightmare beliefs. Tibetan Buddhism is spreading beyond Zayul into the foothills of Assam.

Still, between the official classes lending them money and the church borrowing it on the instalment system, the slave class generally, who form about one-quarter of the total population, are not left with great possessions. Beyond the clothes they stand up in, a coarse and narrow blanket, and a wooden eating bowl, they possess nothing. They live in wooden kennels ranged round the yard of the rich man's house. Sometimes they share a kennel with a sick calf, or a couple of pigs... they don't seem to mind.

Besides the rich Tibetan colonists, and the monks and the slaves, there are other elements in the population. Kampas, for instance. These people come from eastern Tibet, and are the most intelligent, progressive and wealthy of all the Tibetan tribes. Chamdo, the capital of Eastern Tibet, almost co-equal with Lhasa, is the great Kampa stronghold. But Kam covers many thousands of square miles, and contains a million inhabitants.

For the rest there are odds and ends of people from anywhere within the four weeks' radius. Zayul is well south of the main road to Lhasa, though the highway between India and China passes through Rima. It is separated from interior Tibet by a range of snow mountains, and has always been regarded as sanctuary for people who do not care to show their faces in polite society and have forgotten their names. The remittance man, the erring son, the defaulting priest, or their Tibetan equivalents are common objects of the countryside.

We had evidence of this when forwarding unaccompanied baggage by coolies to the next village. Some twenty loads were sent on a day in advance. Two days later at Giwang I checked over the boxes. A 60-lb. case of stores was missing. We hunted high, we hunted low; it was nowhere to be found.

I was going out of the house one morning when Tsumbi stopped me. 'A messenger from Sadiya is arriving, *sahib*. He has letters for you, and a box.'

'Give me the letters, then.'

'He has not arrived yet. He will come to-morrow, maybe to-night.'

'How do you know?'

'People have seen him on the road. They tell me he is close at hand.'

News travels fast in Tibet, and a messenger must travel swiftly if he is to outstrip it. Well, no one expects a Mishmi runner to travel fast; he might not arrive for two days. But when four days had passed, and still he had not arrived, I grew sceptical, then indifferent. We had had no letters for six weeks—not a very long time, but then my hopes had been aroused and I wanted to hear who had won the boat race. It had rained for two days, and Tsumbi said the Mishmi would not travel while it rained, though why he should not was a mystery—a Mishmi should be waterproof. Anyhow, it was fine again now. But no Mishmi arrived. Evidently the story was a myth. I told Tsumbi so; there never had been a Mishmi runner I said, or a mail, or a box of silver. The whole thing had been invented. But Tsumbi shook his head.

'Who would invent such a thing, *sahib*? Besides, Wunju saw him. He would not lie.' And certainly the story was too circumstantial to be dismissed lightly. 'Well,' I said, 'you had better ride back and make inquiries. Tell the lazy fellow to hurry up. Perhaps he is down with fever. Bring the letters. Also look for the lost box of stores. It must have got left behind at the last village.'

So Tsumbi departed on one of the shaggy little ponies, with a great jingling of bells. It would take him a day or two. But obviously, once out of my sight, he would not hurry—he could easily say it had taken him four days, and get away with it. This was the kind of job he loved: and he felt very self-important dashing from village to village on a pony, hobnobbing with headmen, and waited on and looked up to by the simple peasants. Was he not a Babu, a clerk to a *sahib* who travelled under the protection of two governments? I expected Tsumbi back in—well, four days, with my letters. It was a fortnight before I saw him again...

The first night we slept at Giwang it rained steadily. We awoke in the small hours, moved our beds so as to dodge the streams of water coming through the roof, cursed inwardly, and fell asleep again. The carpenters started on structural alterations next day, moved boards about like chessmen, and said it would be all right. The rain continued. Ronald suspended a waterproof sheet over his head to catch the drips, and I re-orientated my bed; all I wanted was a dry spot for my pillow; I dislike being awakened by water on the brain. About four o'clock in the morning a bellow rang through the house. Ronald had kept dry all night, but in the end his ingenuity was his undoing. Before the sun rose his waterproof sheet sagged and tipped the night's accumulated drainage—about a cisternful of water—straight on to his head, and he awoke with a start. After breakfast I told Tsumbi that we must have a watertight compartment to sleep and work in. So we moved into the chapel, where they also kept the quern. We were able to have a fire here, and were quite cosy. As one Zayul house is very much like another, I will describe this one in more detail, and it can serve as a pattern for all.

To begin with, it is built entirely of timber; that is to say of Pine, which composes the forest all round. It is raised on piles—capped by flat stones to keep out the rats—and has a low-pitched roof and overhanging eaves with broad gables. The tiles are simply warped boards; and since all boards are split from the wood with wedges, and squared with an adze, it is not surprising that the lower slopes of the roof leak. There is only one entrance, through a trap in the floor, reached by a steep, narrow stairway; the trap is closed at night by two heavy flaps, and a great stone placed on them, for there are robbers in Zayul. Coming in from the bright light outside, at first you see nothing. A doorway leads out on to a square veranda, where agricultural operations are carried on. Another door opens into the most important room, occupying the whole middle part of the house. This is the kitchen. A wooden fire burns on a sort of hearth near the centre, and the smoke finds its way out of a hole at the top where the wooden slats have been pulled aside; but not before it has furred everything nearby with an efflorescence of soot. Light comes in through the chimney and two small windows, which can be closed by wooden doors. There is no glass or even paper. At night chips of resinous pine wood, flaring on a stone, supply the only illumination, and an abundance of greasy smoke. The end of the house is partitioned off from the kitchen; part of it is a private chapel. However, the Tibetans are not fanatical, and henceforth we occupied this room, which was dry. Ronald ground our rice when he wanted indoor exercise; the heavy granite stones of the quern made a gentle, purring noise.

One other room deserves notice—a narrow strip down one side of the house, under the eaves. This, the guest-room, was the best in the house from our point of view, because it had sliding panels for windows, or boards which we could knock out, all down one side, admitting plenty of light. But at Giwang this guest-room had a defective roof; so after being twice flooded out in the night we invaded the interior of the house.

Giwang comprises six or eight houses scattered down a flight of rice steps high above the river. Its altitude is a little under 7000 feet. Walk out of the headman's house, and the unfathomed forest rises in front of you. So I started to explore. A path led up a stream, through woods blue with Irises; this stream brought water to the paddy fields. Following it up I found myself climbing steeply to where a flume girdled the hill. I followed the racing arrow of water round the steep bulge of rock into the cool temperate forest. The snow had melted, and the trees were thrilling to the stir of life. A fairy wand of sunlight had touched them, and they had answered, throwing off their winter covering with a delicious crepitation of falling scales. Maple trees slowly uncrumpled the most delicate apple-green fans and spread them without a crease. Rhododendrons fifty feet high, which had never lost their leaves, but had twisted them into cigars and tucked them tightly against the branches, had burst into salvos of crimson, primrose and amethyst blossom. The leafless branches of the Magnolias frothed up in a lather of ivory whiteness, each flower as large as a water lily, till the grey trees were transfigured. Presently I reached the end of the flume, which drew its water from a pool at the foot of a fine cascade 2000 feet above the village.

When I got back I told Brooks-Carrington about the place, and next day we went up to shoot the Magnolias. By the time we got there,

MAGNOLIA CAMPBELLII

IRIS WATTII

however, the sun had left the glen; it was a morning shot. We tried again next day; but the sun refused to come out, and we retired baffled. Then, on a brilliant morning, Brooks-Carrington started once more for the cascade; while Ronald and I climbed the ridge towards a pink-flowered tree we had seen from the house. It was about 1500 feet above us, and only difficult to reach because once in the forest we could no longer see it. However, we did reach it; and it turned out to be a solitary pink *Magnolia Campbellii,* of a richer colour than any I had ever seen before. In Sikkim *M. Campbellii* is generally pink; it is the white-flowered form which is rare. When we got back to the village, Brooks-Carrington had not returned. He came in later, hot and indignant.

'Well,' I said, 'you must have got a really good picture at last; you couldn't have had a better day.'

'No,' he replied gloomily, 'but I could have had better coolies. They never turned up with the camera at all!'

I was dumbfounded. But it was true; the coolies carrying the camera and tripod had wandered off in another direction while Brooks-Carrington had sat in the forest waiting. He watched the sun slide over the hill, leaving the great white flowers which a few minutes before had been lambent like porcelain, quite cold; then came away empty-handed for the third time.

'We must stop here until we *do* get it,' I said, obstinately. I had intended to move on up the valley, but I cancelled the order.

The next morning, April 28th, was sunny. For the fourth time we climbed up to the waterfall, and obtained some fine shots. On the way back we found a tree of scarlet *Rhododendron arboreum,* whose fallen corollas strewed the ground like red-hot cinders. As another one drifted down, you almost expected to hear it sizzle and go out, but it remained glowing as brightly as before.

Spring had now definitely come to the valley, although when it rained the temperature hardly reached 55 deg. F. On sunny days, however, it was as warm as England normally is in June. Apple trees were in blossom round the village, filling the air with fragrance: and cuckoos called from the thicket. But if you went into the Pine forest, the silence would presently be divided by a wild screeching, and a flock of brilliant green long-tailed parakeets would dart swiftly by. Cuckoos and parakeets; somehow it seemed an odd mixture, as incongruous as a clump of Bananas and Pine trees I had noticed growing together farther down the valley.

The wheat stood three feet high; but no film of green yet dimmed the mirror of the paddyfields. Looking across the gorge we saw a lot of snow quite low down.

There were many beautiful, and some strange, trees round the villages. A cherry, smelling strongly of almonds, was in full flower. But though there are fruit trees in Tibet, there is no fruit worth eating. Only the walnuts are good. Apples, pears, quinces, peaches—all these grew in the Rong Tö valley; but they might have been wild, so hard is the fruit. As a matter of fact, though, they are quite edible when stewed, and would make excellent jam. They are neither wild, nor are they cultivated; they just grow, like Topsy, in the villages. The only fruits in Tibet which are worth eating are the wild strawberries and raspberries.

Then there was the tree-of-heaven (Ailanthus), complacently rousing itself to put on a new suit of date-gold foliage; and a handsome shade tree, with dragon-green palmate leaves like a maple, called *Schefflera shweliensis*. A film of fragrance enveloped, like a visible haze, a small fluffy-flowered tree called Symplocos. Besides these more exotic species, there were many familiar, almost English, trees in the forest, such as Hornbeam, Birch and Willow. But now that spring had come, I missed that most lovely and most English of all trees, the Beech. There are no Beeches in Asia.

It was time to move on to the next halting place. Solé, reputed to be a wealthy village, was two stages—but only fourteen miles—distant; I proposed to spend ten days or a fortnight there, exploring the forested hills of the frontier and making a picture. So we packed up and moved north again on May 2nd.

CHAPTER IV

FIRST FLOWERS

At Solé we found the river tranquil for the first time since we left Rima. It was flowing swiftly, yet smoothly, in a comparatively wide, flat valley, and there were ricefields right down to the river bank. But the granite mountains behind the village were precipitous. The Pine forest was carpeted with dwarf purple Irises, and flecked with the deliciously scented *Stellera chamaejasme*, each clump of which throws up a sheaf of green-jacketed stems, each stem capped by a puff of chalk-white flowers. The star-shaped flower is indeed white inside, but outside it may be purple, gamboge, or cinnabar-red; so that the flowers twinkle in a breeze, with quick-changing colours, like a rising constellation. It is well-named Stellera, the starry one.

On my first climb up the Pine-clad cliffs which backed the straggling village, I found the seed capsule of a Nomocharis, as I thought; this year's plants were not even showing through the brown grass, so all I had to go by was last year's lily-like capsule, with—heaven be praised!—a little seed still trapped in it. A Nomocharis from 7000 feet altitude in a dry Pine forest ought to be something new; they are mostly high alpine plants, and usually prefer a wetter climate. Later I had reason to believe that this was not a Nomocharis, but a dwarf or starved form of the Tsangpo Lily, or Pink Martagon (*Lilium Wardii*); I collected it in flower higher up the valley in July. In a deep, shady gully which lined the steep face of the mountain was the beautiful nutmeg-scented *Rhododendron megacalyx*, its stainless marble-white trumpets now fully expanded. *R. megacalyx* grows on granite cliffs and rocks, never on trees, and is always found mixed with other shrubs, besides Rhododendrons, often in thickets so tight that it is impossible to break through them. There also grew on these slab cliffs a striking form of *R. bullatum*, its straggling branches tipped with deep carmine buds expanding into flowers definitely flushed or banded with rose-purple. In the much moister Mishmi hills to the west, *R. bullatum* is a common

FIRST FLOWERS

shrub, growing aloft, perched up on the shoulders of the largest trees in the rain forest.

But here it never risked its life in a tree top, and for a good reason; there was no holding for it. In the rain forest of the Mishmi hills every big tree is padded with moss, great sponge-like growths, which afford anchorage for a variety of plants; but in this drier air there was no moss bed. And so *R. bullatum,* like *R. megacalyx,* clung to the cliffs, where it had ample support. In the Mishmi hills, the flowers are dazzling white; but the cleaner, drier air of Tibet had brought colour to those pale cheeks, and fragrance also.

One more flower I will mention before I leave this fascinating topic: the daintiest little cream and snuff coloured slipper orchid (Cypripedium) which grew on the steep Pine-clad slopes. It was so tiny, I almost overlooked it; yet it was a common gem. One does not associate the exotic Orchid family with Tibet, but there are many kinds of small ground Orchids, some of them deliciously scented. The dry air of High Asia distills wonderful aromatic scents not only from the flowers, but from the leaves and seeds of plants.

One morning I was surprised to see three queer dwarf men in the yard, where the paddy is husked in wooden mortars, and pigs are fed, and cattle milked. They were sitting on a dunghill, weaving bamboo baskets. Their hair was cropped and they wore cane helmets on their heads; otherwise they were dressed like Digaru Mishmis; that is to say, they wore a sleeveless cloth jacket and a sporan-like flap of cloth, elaborately decorated, suspended from a string round the waist. Each carried two knives, and they smoked incessantly out of long silver pipes like converted candlesticks. Two of them looked suspicious and surly, but the third, an aged headman with a well-creased mahogany face like a carved Buddha, smiled amiably. They came from the unexplored Tangon river over the snow-covered range to the west, and belonged to the Bebejiya clan. There are several passes to the Tangon river, one a few miles south of Solé, another just north of the village, a third above Mugu. It was by the pass north of Solé that these jungle men, driven by hunger, had crossed the mountains, before the snow closed the pass. When the snow melted, in July, they would return to their homes in the jungle. They might well stare at us: they could hardly have expected to see three specimens of the redoubtable white race of whom they had heard vaguely, though never seen, in Tibet. This is the only pass between the Dihang valley and the Rong Tö which is used; and that only by the Bebejiyas. Many years ago, Wunju, headman of Giwang, while hunting musk deer, had crossed a pass to the south, descended to the Tangon river, and reached Sadiya in nine days. A native will travel at least twice as fast as a white man; so we may well allow three weeks from here to Sadiya by that route. But I strongly suspect that few white men would do it in that time. In the winter I climbed up to the pass, and looked over into a deep valley, which was the source of the Tangon river, a tributary of the Dibang. But the pass had been 'closed' by the monks; so I did not descend into that alien land. When the lamas close a pass, it is understood that anyone who crosses it does so at his peril. As a matter of fact, no one ever does. The men of Zayul go up to the pass from their side, the Bebejiya do the same on their side. But that is all: it is as though there was no pass. The reason which prompts this step is, of course, feuds. One year a Bebejiya hunter shoots a Zayul hunter in the mountains: perhaps accidentally, perhaps not. Nor does it matter on which side of the mountain it happened: the important thing is, it did happen, and in the eyes of the Zayul hunters there can be no extenuating circumstances. An eye for an eye. In the following year a Zayul hunter shoots a Bebejiya hunter—possibly accidentally, probably not. The score is now one all. But the Bebejiya victim is not closely related to the Bebejiya who successfully shot a Tibetan the year before; and his family feel that they must avenge his slaughter, or else for ever walk ashamed. So they murder a Tibetan, and the Bebejiya are one up. The feud is now well started on its murky way. At last the church intervenes on the Zayul side, as representing the most highly civilized state. The lamas decree the pass closed, as the headmaster might decree an undesirable place out of bounds. It is used no more; and traffic, if any, is diverted to another pass. It is forbidden to take life in Tibet proper, especially near a monastery. But in the forested provinces and districts of the southeast, many villages live largely by hunting. There are plenty of hunters in Zayul. Solé is not a hunting village; but the inhabitants of Giwang and other villages across the river are keen hunters.

The animals commonly hunted are musk deer, *gooral* and takin (*Budorcas taxicolor*). Musk deer are sought only for musk, which is sold in Sadiya or Darjeeling. The price of the pods, more or less dry, in Zayul hardly varies, being seven Indian rupees a tola—that is to say, seven times its weight in rupee silver. But in the Indian bazaars the price fluctuates widely, from about eight rupees to twenty or more a tola, according to supply and demand. Gooral and takin are both hunted for their skins, used to make jackets, caps, flour-bags, and occasionally trousers; and also for meat. The black Himalayan bear (*Ursus torquatus*) is also found here. The only animals I saw at this time were a large sandy-coloured marten (*Charronia flavigula*) and a number of squirrels.

We now decided to assemble the collapsible boat which I had brought to navigate the lakes; it might also, I thought, prove useful on the return journey. Here was the first stretch of water we had seen since leaving Sadiya which was navigable, at any rate, to the unskilled boatman. Ronald, who had boated on the Danube, proved invaluable in putting the thing together. It seemed to have either too many parts or too few—I couldn't be quite sure which. We carried it down to the river in its bags and assembled it in the presence of an admiring crowd on the shore. I felt rather in the way and confined myself to superintending operations while Ronald sweated with frames, stretchers and skins as he lay on his stomach, his head below hatches. Finally he emerged, hot and red, from the bowels of the finished article.

'At any rate,' I thought, 'Ronald, having assembled it, leaving no spare parts over, will want to have the honour of embarking in it first. He has earned the right; it would be churlish of me to deny him, and insist on the leader taking the first plunge.' (The river flowed swiftly by, effervescing; it was cold glacier water, very clear.) I felt almost heroic, in the sacrifice I was about to make—but it would be better to register reluctance, I thought, and then acquiesce gracefully.

Ronald picked up the canoe and plumped it into the water by a rock; he held on tight against the tugging current. 'Now, Chief,' he said amiably.

My heart stopped. So that was that: and I hadn't boated for years! Very well... I stepped gingerly into the wobbly canoe, seized the paddle, and was pushed out into the racing river. But the canoe was as buoyant as a cork, and I paddled across easily enough, disembarked fifty yards downstream, and paddled back again. Confidence had returned, and I was ready to start shooting rapids in a collapsible canoe which, so long as it did not collapse at the wrong moment, was probably unsinkable. Ronald then took a turn. We launched the craft once more while we were at Solé. It was the last bit of navigable water we saw; and the collapsible canoe was dismantled for the journey to Ata—and never expanded again.

On May 12th, I decided that it was time to go for a high climb and reach the temperate forest if possible. There was a lot of snow above 10,000 feet, but below at lower altitudes the buds should be bursting on the trees. The cliffs above the village, as far as the ridge, were clothed with Pine; beyond that we could not see what happened. I must reach the ridge. So I started off, carrying a rucksack, and at the end of half an hour had ascended 833 feet. Another hour's climbing, and I was 2019 feet above the village; still no change in the vegetation, but I saw some beautiful rosy-budded, though straggly, plants of *Rhododendron bullatum* clasping a rock. Now climbing became much harder; I was approaching the ridge, and the cliffs were precipitous. I went cautiously, looking for the easiest way up. At one o'clock, after nearly three hours' climbing, I flung myself down on the ridge, to get my breath and enjoy the magnificent view down the valley. I had climbed 3275 feet and reached my first objective; but could I get back? However, I was too interested in what lay ahead to bother about that now. It was a glorious, sunny day, I was nearly 10,000 feet above sea-level, and I had no appointment to keep. So I wandered on up the ridge, where it was easy going except for thickets of bamboo grass. The forest came right up to the brow of the ridge from the far side. Here grew a variety of trees, including *Tsuga yunnanensis* (the only Conifer), *Rhododendron sinograndi,* and *R. hylaeum, Acer Wardii, Viburnum Wardii,* and two species of Litsaea. The ridge became steeper, and the going more difficult; forest had invaded the ridge itself, and was beginning to creep down the slope on our side. I was scrambling up big moss-covered rocks, amongst patches of snow and sombre-looking Fir trees; the forest had changed; there were Larches, with no hint of green on their stubbly branches, and it was suddenly cold. At 2.15, having been climbing four hours, for a gain of 4298 feet, I turned back; and reaching the sunshine, I threw off my rucksack and ate my lunch ration—two chupatties and a bar of chocolate.

I had passed 10,000 feet, and really there was not very much in flower in the forest; the alpine region still slept soundly under its deep quilt of snow. But the flowering trees were beautiful, none more so than my Viburnum, wreathed in white blossom, with its crinkly deep green leaves already fully developed and shining in the sun. As for the

FIRST FLOWERS

Litsaeas—small trees not seen in England—they were heavily beaded with canary-yellow drops, the glistening silver tips of the leaves just peeping out of dark chrysalis buds. My red-leafed Maple was opening shyly thin parasols which would presently expand into pronged leaves; flowers swung from every branch.

At 3.30 I started down the steep face; I did not remember exactly where I had come up and almost immediately I found myself in difficulties. Granite rock, on an exposed slope, has an awkward way of weathering into vertical slabs on which nothing will grow; they crop out suddenly and unexpectedly, converting a steep enough slope into a nasty precipice, with a drop of thirty or forty feet. It is not so bad when ascending; but when descending, with the pitch below hidden because it bulges as it steepens, and your rucksack catching on the rocks threatening to trip you up, unpleasant moments arrive. Once I clung trembling to a ledge, unable to move up or down...there were bright yellow Violets growing here on the Pine-scented slope, and I looked at them fixedly to ease my mind ...presently I recovered and got down. Eventually I reached the village with a good collection of specimens, after nearly eight hours' climbing. It had been a full day and successful.

Tsumbi met me. He looked grave.

'My letters, Tsumbi?'

He handed them to me without a word. They were dirty.

'Well?' Tsumbi had been away a fortnight. I was waiting for some explanation.

'There has been a dreadful crime, *sahib*. The Mishmi mail runner was—murdered; in the forest, not far from Giwang. The letters have only just been recovered from under a rock.'

For a moment the seriousness of the position hardly dawned upon me. I was startled. When it did I asked Tsumbi to tell me all he knew. It was not much. The Mishmi had arrived within a day's march of Giwang, just as Tsumbi had told me. He had slept at a village that night and started off very early the following morning. In a lonely part of the forest he had been attacked; two great knife-cuts had split his head and chest open. The corpse had been flung into a ravine; it was several days before his body was discovered, more days before the letters were found under a stone. The murderer had left them—or some of them; we shall never know how many were lost...

Obviously the motive was robbery: everyone knew the box he was carrying contained money.

'Tsumbi,' I said, 'this is a very serious matter. That Mishmi was sent to me by the representative of my Government. I am a guest of the Tibetan Government. No sooner does my messenger cross the frontier than he is murdered.'

'*Sahib*, the Governor begs that you will go and see him. He is very distressed. He kept saying, "What will the *sahib* say when he hears of this?" '

'I will go at once,' I said.

I determined to start first thing the next morning, and ordered ponies to be ready. I would travel light and travel fast. I communicated this decision to my companions, and told them to go on to Rongyul, the next village up the valley, and await me there; I expected to be absent eight days.

CHAPTER V

ORDEAL BY FLOGGING

It was a gloriously clear morning when Tsumbi and I rode out of Solé with one baggage pony and two coolies.

Glancing back up the valley, I caught sight of a fine peak far to the north; it was my first sight of the snow range. We travelled fast, and by evening had covered three of the official stages. Starting again early next morning, we rode all day, and at dusk reached the rope bridge opposite Rima. A storm had sprung up, and a boisterous wind was tearing the clouds to ribands and flinging the rain in our faces. We reached the village after dark, wet through. No sooner had I changed and sat down than the Governor of Zayul was announced. I greeted him warmly; we did not however discuss the murder, and after a few polite inquiries he left. At ten o'clock that night I got my mail, which had just arrived at Rima. Dawa Tsering had sent a messenger to Sadiya on April 11th with our letters; the man got back to Rima on May 15th, five weeks later, which was good going. This was the last communication I received from the outside world till I reached Assam just before the new year.

Rima was warming up; most of the snow had melted on the surrounding mountains, and after the cool air of the upper valley it seemed hot and sticky. Early next morning I called on the Governor, and found him busy reading through a bundle of statements made by the various headmen in the Rong Tö valley. There was an air of purpose about him which I liked. To my question, whether it was known who committed the murder, he said no, he was still making inquiries; but most of the men summoned to Rima had arrived. The Governor, I learnt, had taken prompt and vigorous action. All roads were closed and no one was allowed through without a pass. The murderer, if he was still in the valley, could not escape, for the mountains were deep under snow.

About midday the official proceedings began. All the headmen from up the valley had been summoned and several villagers also;

I was sitting with the Governor in his small room drinking tea when two lesser officers arrived. Then, one by one, certain men were brought in and questioned as to their movements on the day of the murder—which, somewhat to my surprise, seemed to be generally known. It was the day after Tsumbi had mentioned the imminent arrival of the runner; yet a fortnight had elapsed before I was informed of the crime. One of the first persons questioned was the headman of Dri—the village mainly responsible for the theft of the stores box, as it turned out. All three Tibetan officers asked him questions, to which he replied. His replies were not recorded, but I gathered they were unsatisfactory. A Mishmi came next. He also stood up to a barrage of questions; I remarked a note of anxiety in his voice, a faint protest as it were. The Governor was trying to establish who had last seen the murdered courier alive, who had first reported the crime, and who were in the neighbourhood on the fatal day. A lama of Dri, dressed in a rich red robe, with a purple wrap folded round his body, stood before the tribunal, and was cross-questioned. The junior officials talked amongst themselves, and somebody laughed. One of the Governor's bodyguard—I nicknamed him the First Lictor—a big man six feet tall, came in, whispered something in the Governor's ear, and retired.

The Governor turned to me, and spoke blandly. 'Shall we go outside?'

We went out on to the low veranda which ran round three sides of a square, enclosing a piece of flat, dusty earth. A prayer flag hung limply from a leaning pole in the centre; just beyond stood a crowd of Tibetan men and women, with a few children. I noticed several Mishmis amongst them. Two lictors leaned carelessly against the pole; they had dropped their *chupas* from their shoulders, tying the sleeves round their waists, which bulged. Each wore a thinner garment, something like a shirt, beneath, and they had rolled up their sleeves, baring brawny arms. In the dust lay a length of rope and two long rawhide horsewhips with cane handles.

A wooden bench was brought out. 'Please be seated,' said the Governor, and we both sat down on the bench facing the crowd. Tsumbi stood at my elbow, and the two minor officials stood up behind the Governor. I looked across at the mountains, silver-crested with snow, and down the valley, which twisted away into the reeking clouds; but over Rima the sun shone in a blue Tibetan sky. An immense silence

GECHI GOMPA

hung over Zayul; here was drama. The Governor nodded and a man stepped forward; he was one of those who had been questioned inside. He knelt in the dust at the Governor's feet, clasped his hands before him, and waited with bowed head.

'They say you were the last person to see the Mishmi alive?'

'No, Kushog.'

'Did you not meet him on the road near Dri?'

'No.'

'But you reported that he had been murdered?'

'It is not true, Kushog.'

'Lobsang says it is true.'

The questioning went on for a minute or two, and then the Governor signed to the lictors, who advanced.

The Governor turned to me. 'Do you object to this man being flogged?'

'Is that the custom?'

'Certainly. He is not telling the truth.'

'I do not wish to interfere in this matter. Please act according to the custom of your country.'

Tsumbi then whispered to me, 'How many strokes?'

I shook my head. It was not my business to say anything.

'A hundred to-day,' said the Governor suavely, 'and a hundred each day until they speak the truth.'

Amidst a deathly silence the untruthful witness lay flat on his stomach. A man squatted cross-legged on the ground and rested the witness's head on his lap, clasping him tightly round the neck in order to keep it there. At the same time a very ragged white-haired man, with senile deliberation passed the rope, doubled, twice round his ankles, drew the two ends through the loop, put his foot on the slack, and holding the ends over his shoulder, turned his back on the witness, and pulled as though hauling a log. There was no doubt that the man who lay in the dust could not move; he was held in a vice. The two big men picked up the whips, and took position, one on either side... the crowd pressed forward but made no sound: a small curly-haired Tibetan dog ran out and barked, till a woman grabbed it. One of the lictors pulled the witness's *chupa* up over his body, baring his buttocks. The Governor nodded...Thwack!...thwack! The whips, wielded by powerful arms, whistled through the air and fell with a sharp crack quickly one after the other. As the first whip was withdrawn and raised slowly for the next stroke the second man struck and tolled the score: 'One!' Thwack...thwack... 'Two!'...'Three!' droned the scorer.

At the first blow, the witness cried out; then, as the raw hide thongs rose and fell, raising dead-white welts on the dirty flesh, and cutting parallel lines in the dust as they trailed upwards, the man broke into groans, succeeded by sobbing and moaning. The crowd—there were thirty or forty people now—looked on impassively; except for the monotonous crack of the whips, and the calling of the count, there was no sound but the wild moaning of the witness...'Twenty!' The Governor held up his hand. Still lying in the dust with his feet bound, the witness's head was released, and he was bombarded with questions. He answered shakily; but the answers were not considered satisfactory, and another twenty blows were dealt. He had ceased to sob now; he prayed, a single prayer, uttered rapidly in a high-pitched voice, repeated again and again. What was it? 'My God! My God!'...Another twenty strokes, and the questioning was repeated, but to no purpose: the full sentence was inflicted...'Forty-eight'... 'forty-nine'... 'fifty,' and the lictors tossed their whips aside. The witness, having received his hundred stripes, was now released to crawl away to a corner helped by a friend who poured white of egg on to the wounds. No blood had been drawn, which struck me as curious.

It was now the Mishmi's turn. He did not kneel, nor did he clasp his hands as he answered the Governor's questions defiantly. But for all his bravado, there was a note of anxiety in his voice. Evidently his replies were not satisfactory and the Governor shook his head.

I asked what he was saying, and Tsumbi explained that the Dri man, who had just been flogged, said the Mishmi had told him he met a man named Lobsang on the day of the murder; the Mishmi denied meeting Lobsang, and said that the Dri man had told him *he* had met Lobsang. 'One of them must be lying,' added Tsumbi, thoughtfully. 'Or both of them,' I added.

Then the Mishmi was stretched out on the ground, and before ever the first lash was lifted he began to scream in a high-pitched voice. 'So much for Mishmi stoicism,' I thought. 'The Tibetan did not make half as much fuss.' All the time he was being flogged he yelled and screamed like a child, suggesting a sort of outraged astonishment, as though a malignant spirit had taken advantage of him. But the flogging went on to the bitter end, till he too was released, and crawled away a blubbering cripple.

Then it was that a tall, sinister-looking Tibetan knelt before the Governor with his hands clasped before him, and his head bowed to the dust. He was not pre-possessing to look at; his complexion was swarthy and he had a hard face—there was something ruthless in his expression and in his steady dark eyes, the sort of cold fury I have seen in a snake's expression. He answered the questions put to him in a low, even voice, and gave some account of his movements on the 'third day of the fourth moon in the year of the water monkey'.

'Who is he?' I whispered to Tsumbi.

'Lobsang, the man suspected of the murder. The courier slept in his house the night before...'

That explained a good deal. It was over this man that there had been disagreement and denial amongst the witnesses. Two had been flogged already...I began to take an interest in this hard-faced suspect. To begin with, he was well, even showily, dressed. He wore long scarlet-topped cloth boots and a blue CHUPA of good cloth. A large silver ear-ring decorated with a turquoise hung from his left ear, and his hair was curly—no uncommon thing in eastern Tibet. Obviously he was not a common coolie, but a man of substance. Afterwards I learnt something of his history. This was not the first time he had been accused of murder; he was known to be a ruffian.

To some of the questions put to him he gave no reply. While he still kneeled before the Presence, a fourth witness was called. He had spoken with two voices, and was duly flogged. Lobsang never moved a muscle, though the space round him seemed to be filled with cripples, who crawled and moaned, sat up, and lay still on their faces again.

Now it was the lama's turn. I was surprised to see him kneel in the Presence, like any coolie. The questioning continued, and there was a terrible fear in his face and voice as he replied. I saw beads of perspiration break out on his forehead; his hands were clasped in supplication, and his black locks stood out straight from his head. He had seen three men flogged, and was in an agony of apprehension; possibly also his dignity was injured at finding himself in such a position. Still I felt sure the Governor would not dare to flog him. But there I was wrong. Clearly he was not satisfied with the lama's answers, and though someone raised his voice in feeble protest, the Governor paid no attention. Instead he turned to me, asking whether I agreed to his being flogged. This time I felt justified in voicing my opinion.

'Why? What does he say?'

'He denies that he saw either the courier or Lobsang on the day of the murder. He says he knows nothing about it.'

'Then please do not beat him.' There was no reason to think that he was either lying or withholding information; and the Governor agreed that he had not contradicted any other witness.

So the lama was released. He thanked the Governor profusely, swearing eternal fealty. His enormous relief at his escape was manifest; the startled, hunted look left his eyes, his skin assumed its normal colour, his long hair fell back into place. As for the Governor, though I do not think he would have hesitated to flog the lama, had he believed it to be his duty, no doubt he was glad to reprieve him.

That concluded the day's proceedings, and I retired to a large Tibetan tent which the Governor had pitched for me.

It is to be noted that, though I had been present at the opening of an inquiry into a murder—not at the first day of a murder trial, there was little difference. A sort of coroner's inquest, without the deceased, and trial of a suspect by a process of elimination, were going on simultaneously. The whole inquiry would be conducted on these lines. Obviously no one had seen the murder committed. I observed that it had not been reported at all really; news of it had seeped out several days afterwards. What then of circumstantial evidence, the cumulative facts all pointing to one man, which is the evidence on which, even in England, most murderers are hanged? What did the murderer do after the crime? Where did he go? What of his clothes, his bloody sword, the heavy box? Who in the neighbourhood had suddenly become rich?—an obvious line of inquiry for future use. Still one could hardly expect any very subtle methods of arriving at the truth. Tibet is a crude country: its civilization is equally crude. The method adopted to unmask the murderer might savour of third degree; but the people knew no other, and were well accustomed to this. Equally interesting was the fact that there was no defence, beyond passionate denial on the part of accused. Once the magistrate had fixed on a man by the process of elimination, that man's chance of escape was a poor one. It was not a trial as we understand the word: rather was it trial and error. At least there was no chance of the murderer, if murderer he was, getting off on a point of law.

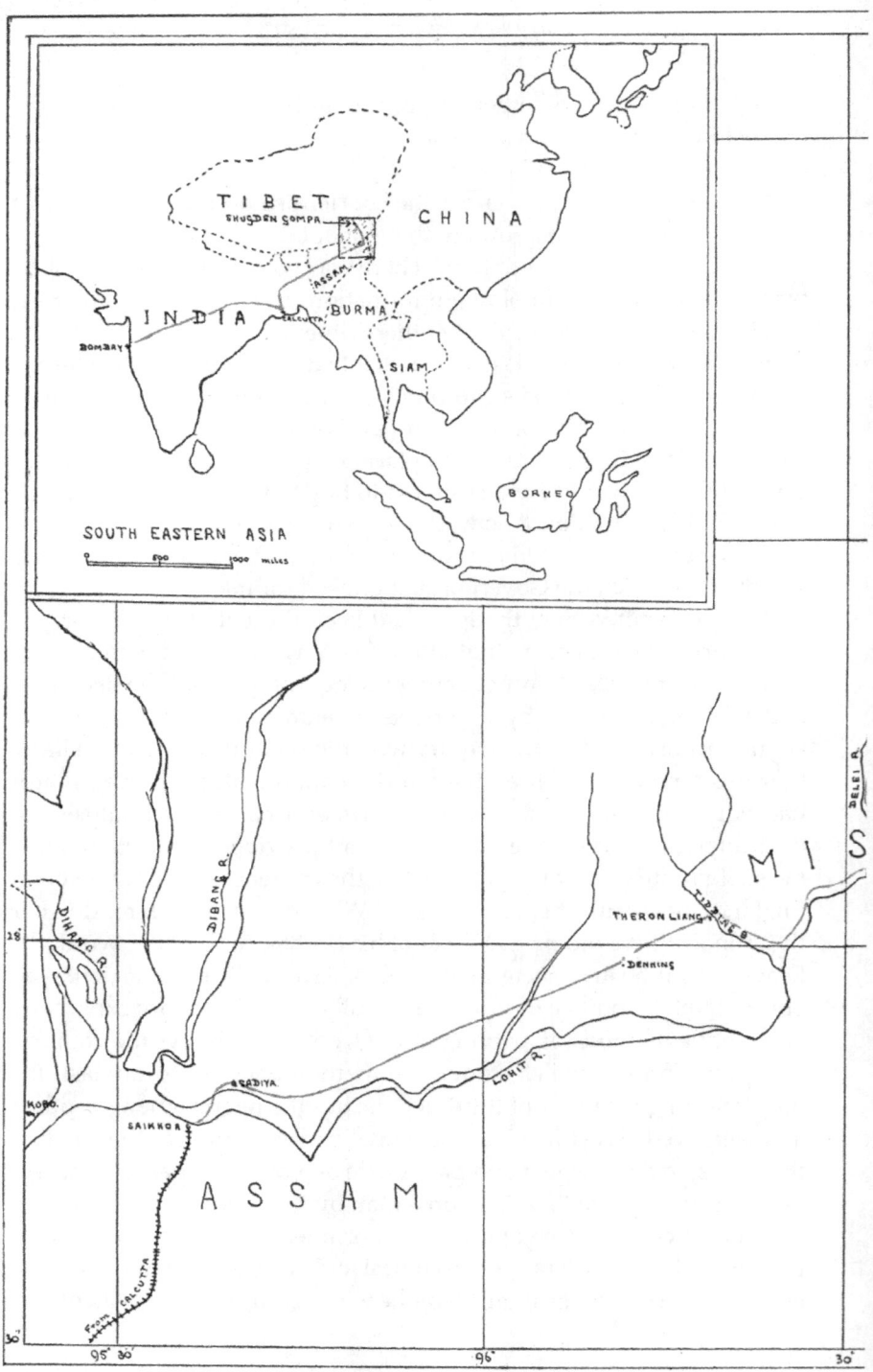

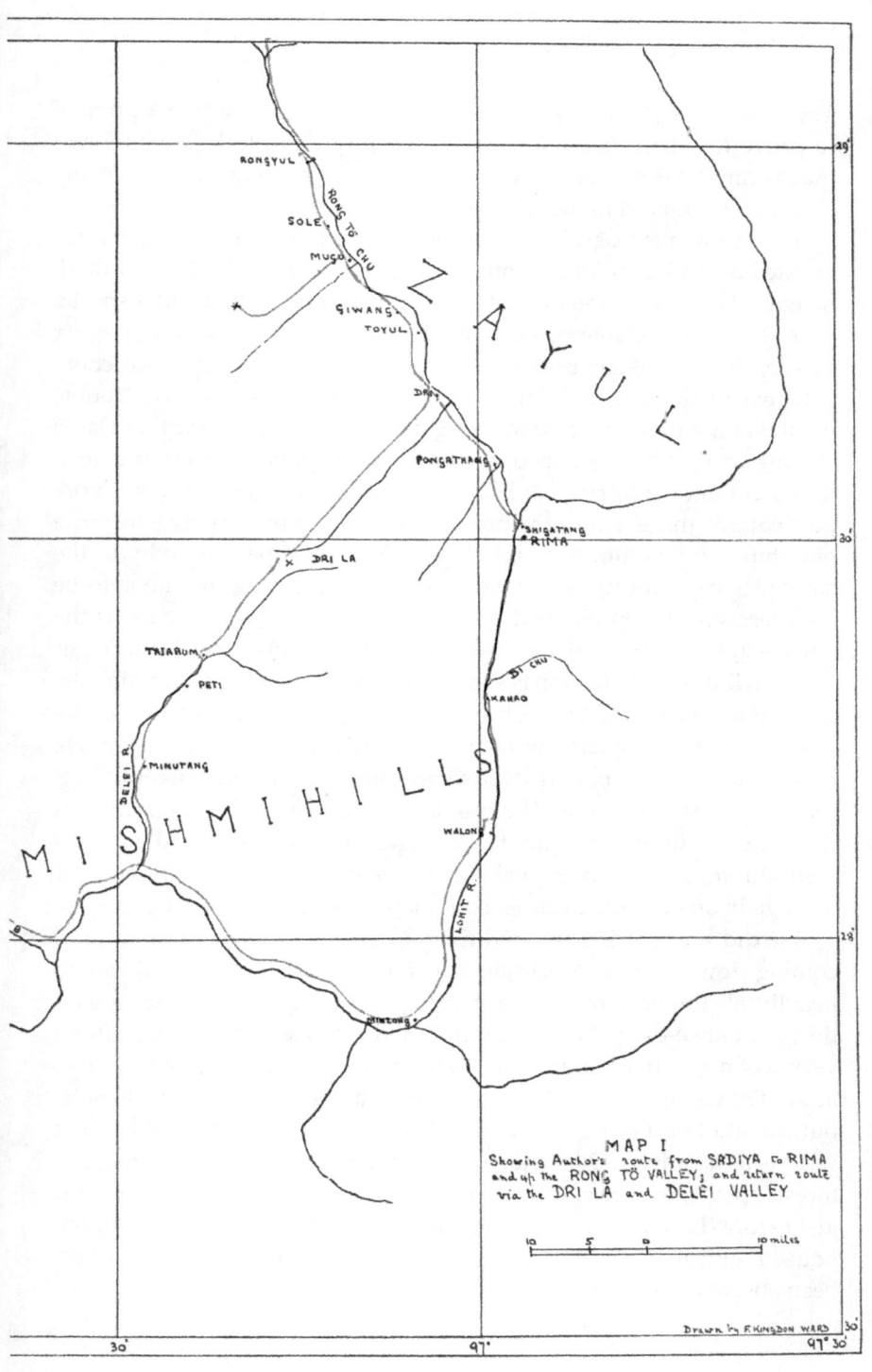

MAP I
Showing Author's route from SADIYA to RIMA and up the RONG TÖ VALLEY, and return route via the DRI LA and DELEI VALLEY

Drawn by F. KINGDON WARD

There could be no forensic triumph at Rima; no persuasive argument to prove to a hard-hearted, soft-headed jury that black is white, and that a criminal is not really a criminal but a much maligned and worthy citizen, without a stain upon his character.

I left Rima next day. The Governor pressed me to stay longer, but I insisted on returning to my companions, as we had work to do. I think he was glad I had come to see him. Before I left, he wrote out a special pass for my companions, authorizing them to accompany me as far as Ata, the last village on the road to Shugden under his jurisdiction. Wherever I showed this document to the headmen, I had no trouble whatever in buying food or in hiring transport. Then I knew that Dawa Tsering really was my friend, for he had not spoken with two voices. It is a common practice, both in China and Tibet, for officials—more particularly those who care not to shoulder responsibility—to write one thing and to cancel it verbally. Wishing to appear friendly to the stranger, they send written instructions to the people that he is to be well received, given all facilities, and generally helped. Happy in the knowledge that the official instructions will be obeyed, the stranger departs. But his satisfaction is short-lived. Arrived at his destination, he finds himself baulked by endless delays. He cannot understand it. The reason is that the official, outwardly friendly, was secretly hostile. He had written fair words, and at the same time had sent verbal instructions that they meant nothing. The message is on record: the cancelling is not. If by any unlikely chance the stranger complains—well, the people were stupid, and will be punished, of course. But Dawa Tsering did not indulge in any double dealing. So I knew that he was really a big man.

We did not make such good time on the return journey as we had coming down. It was nearly noon before we left Rima, and it rained heavily all the way back. Every village was a quagmire—not ankle-deep, but knee-deep, in black mud. When I awoke on May 18th, after a very wet night, fresh snow powdered the Fir trees a few thousand feet above Dri village. On the third day, in drenching rain, we rode into Solé: but Ronald and Brooks-Carrington had gone on to Rongyul, the next village, as arranged. Once more we plunged into the dripping, gloomy forest, splashing through the pools, wading the swollen streams, till just before dusk we reached Rongyul. I found my friends comfortably housed, and in the evening we sat down to a pleasant dinner. I had been absent only six days.

The end of the story of the murdered courier, so far as I know it, is as follows. While at Shugden Gompa, I received a letter from Dawa Tsering, saying that they had found the murderer. They had not recovered the money, but if I needed it, the local government would refund it. I, of course, waived the money, in consideration of the trouble the local government had been put to; and on my return to Assam I compensated the victim's family. So much for the financial side of the affair. It seemed that Lobsang, the suspect, was the murderer after all—a conclusion not in accordance with the best detective traditions. When I left Zayul, in December, he was a prisoner at Sangachu Dzong, waiting to be sent to Chamdo, for the Governor had no powers of life and death. His ultimate fate was uncertain, for the modern Tibetan government, having abandoned the barbarous practice of mutilating criminals in vogue twenty-five years ago, has swung to the other extreme, and is chary of inflicting the death penalty. All the same, I would have put no money on Lobsang's life. It was all Lombard Street to a China orange that he would be executed. As to the missing stores box, that also was eventually discovered by the sleuths of Rima—empty, of course. A coolie, curious as to its contents, had quietly dropped out of the line and taken it home with him. Unluckily for him he had forgotten to take a tin-opener; and I pictured him attacking the tins of Maconachie and condensed milk with a blunt sword, or battering them to pulp on a rock. I wonder what he thought of the contents, and whether he cooked any of the tins? I hope he cooked the chocolate...anyway, the tins were reported missing. As the coolie had no visible means of support, he was not asked to compensate me. Instead, the Governor fined the two villages, between which the box vanished, 150 *tankas* each, about £3 10s., or nearly the net value of the goods. I made no claim to this, and it went to swell the government coffers at Rima. What befell the coolie I was not told; perhaps his friends, who did not share the tins, but shared the fine, attended to him.

CHAPTER VI

THE RIVER OF ICE

My main object now was to reach the alpine region as speedily as possible to see the Rhododendrons in bloom. Rhododendrons flower early, while the snow is yet melting; many flower *in* the snow. By the end of June, the best of them are over. At Rongyul I met a Chinese carpenter who came from Yunnan. He was in very poor circumstances, and lived in a serf's hutch in the courtyard of the headman's house. His clothes were in rags and he lived on the coarsest fare. Nevertheless he had a wife and child, both sadly under-nourished. In the evenings he would come up to my room and talk. Sometime he wanted to return to China, but what was the use! He had no money, and his carpentering did not bring him in much. However, a Chinese craftsman in the Rong Tö valley is certain of a little work, and he was often away on a job, travelling even as far as Rima. He told me that there were Kiutzu slaves in Rongyul, and brought one in to show me. 'Kiutzu' is the Chinese name for a pygmy tribe inhabiting the Irrawaddy jungle. We call them Daru; the Tibetan settlers in the Irrawaddy basin call them Tellu—which is the same word pronounced as a Tibetan would pronounce it. Of the Daru pygmies it may be said that the women, no beauties, tattoo their faces, making them still more hideous. They have long been the victims of Tibetan and Chinese oppression, so that, until quite recently, although working in their clearings, fishing and hunting by day, at night they retired to the tree-tops to sleep, fearing to be on the ground after dark, lest they be surprised and carried off into slavery. There is a regular traffic in slaves in south-eastern Tibet, especially children, and high prices are paid for a likely girl.

The weather now turned very wet and the village became a slough. The street was a foot deep in black slime, and there was one morass in front of the headman's house where you became immersed to your knees in glutinous filth; it was better to avoid the street altogether. No wonder the houses were built on piles, which were set on large flat stones, or sometimes two stones! It was all the foundations there were. Some houses had taken a list to starboard, owing to a pile sinking; others had gently collapsed or sagged like wet cardboard. They were nervous in Rongyul, and the inmates of my home closed the trapdoor at night. I found it impossible to get out after dark without creating a disturbance; and although I might easily have dropped to the ground from my room, the window frames were far too narrow for me to get through. There remained only the large open veranda in front of the house, and as the mastiffs were chained to the supporting pillars beneath, that route also was barred.

The headman of Rongyul was blind and a cripple. He was carried about in a chair, but spent most of his days lying on a heap of skins.

A new rope bridge was woven for us and tied in position, and on May 24th we crossed the river and continued our journey up the valley.

Rongyul is the last village where rice is grown. Henceforward the chief crop south of the great range is maize; but there are few villages north of Rongyul. After marching along the side of the cliff all day, we camped in the Pine forest. Here I almost trod on a snake (*Zaocys nigromarginatus*). The front half of it was bright green, the end half almost black with thin yellow stripes along each side; it was five feet long.

In the Pine forest here I observed great numbers of lily-like plants coming up through the bracken. I again mistook it for a species of Nomocharis, since it bore only one or two flower buds; but later it appeared that it too was a starved form of the pink Martagon Lily, *Lilium Wardii*.

As we approached the mouth of the Ata river the valley broadened out, and a forest of derelict trees, which had perished in a flood, appeared. They stood gaunt and lifeless in the wide, shining, shingly river bed, though no water flowed near them now. Cattle grazed on the bracken-clad slopes.

Marching over the devastated area, we crossed a broad, swift stream by a rickety bamboo bridge, and turned to see whence it came. There was a sharp cleft in the mountains, and the stream came pouring out of it into the broad valley. I turned northwards to follow the main river, when I saw the coolies trudging across to the cliff and disappearing one by one, as it seemed, into the living rock. So this was the Ata river!

I went back to look at the main river above the confluence: that is, the Rong Tö Chu, though from this point it is called the Zayul Ngu chu. It was vastly bigger than the Ata stream, a point on which I had been in doubt up till that moment. So the Rong Tö Chu must rise many days' journey to the north-west.

Returning along the right bank, I followed the others into the cliff from which gushed the stream. Just round the corner was a bridge made of sticks, and we recrossed the formidable Ata stream, inside the gorge, a few hundred yards from where we had first crossed it. Now began a difficult scramble along the wooded cliffs, the gorge growing narrower and the stream ever more turbulent. The rocks echoed the noise of tumbling waters. Presently we reached a suspension bridge made of two ropes of twisted bamboo, with cross-ropes tied to them at intervals and supporting a narrow footway of rough planks. The bridge was shallow, and it swayed so much that I felt all the time as though I might easily be tipped out of it. Imagine walking along a hammock, suspended over a river forty yards wide, and you will have some idea of this bridge. We were buried in this remarkable gorge, with the river twisting away out of sight. Evidently the fall was very steep, for we heard the hollow booming of cataracts round the corner. The cliffs rose almost vertically for thousands of feet; far above a riband of blue sky showed between the trees which leant across to each other. The frantic torrent rushing between blocks of granite, sculptured into the semblance of grotesque statuary, filled its bed completely. How we were to get out of the gorge was a mystery. However, it was quickly solved, and nothing could have been simpler. We climbed up the cliffs by means of ladders, working from ledge to ledge. It was dangerous work—for the coolies, at any rate. One false step and we must have been hurled into the abyss. As for the ladders, they were, after all, only thin treetrunks, with notches cut in them for steps. So shallow were these steps, that one could barely put one's toes in them—certainly they were never designed for people wearing leather boots! And the ladders only leant against the cliff. Hundreds of feet above the torrent, the gorge opened out a little. The side streams, left far behind as the glacier torrent bored its way into the bowels of the earth, leapt the last thousand feet in a splendid endeavour to catch up. They fell in long silver threads looking like drawn wire, or spread out in a curtain of spray. The scenery was piling up. Passing through patches of forest, we came out into a bay of the mountains, and saw across bare stony fields the tiny village of Modung, embowered in trees.

Kyipu, the headman of Modung, is quite a remarkable man. To begin with, he is very rich, owning land, serfs, ponies, many head of cattle, two score pigs, and other forms of wealth. He buys musk—the scent gland of the male musk deer—from the native hunters and sells it in Sadiya. With his profits he has built and endowed a small monastery a short distance above Modung, where the Suku and Ata streams unite. This monastery, called Gechi Gompa, was blessed by the Kalon Lama, who in 1913 came from Lhasa to command the eastern army defending Chamdo against the Chinese. The Kalon Lama, when asked to prophesy where the monastery should be built, wrote 'nam tu sum, ri tu sum, chu tu sum', and the monastery was therefore built where the three valleys unite, under the shadow of high peaks.[1] Thirty monks are attached to it and live in the surrounding villages. The head lama, a boy of twelve, is a reincarnation or living Buddha: I was greatly struck by the benignity of his young face.

After spending a few days at Modung, we marched to Ata, the last village up the valley: indeed a great glacier pushes its snout down to within two miles of the village, though the altitude at its foot is barely 8000 feet. From the time we saw this glacier, and the great snow peaks commanding the head of the valley, we had eyes for nothing else, and our efforts were directed towards getting up into the snow.

The plain truth that a glacier had once filled the entire Rong Tö valley and descended beyond Rima, sixty miles distant, no longer seemed fantastic, now that we had actually seen a glacier. It explained, too, the Ata valley, and the sudden drop through a narrow gorge into the main Rong Tö valley below; clearly the glacier, while slowly retreating, had long remained stationary at some point below Modung, severed from the main glacier below, which was also retreating. Gradually the stream—much bigger than the Rong Tö Chu of to-day—had excavated the main valley till it stood a thousand feet below the end of the Modung glacier: and the stream from the Modung glacier had leapt over the cliff to meet it as the side streams do to this day—but had gradually worn a gorge through the crystalline rocks, as being easier. It is still cutting the gorge; while yet smaller streams, sprung likewise from glaciers, continue to leap the cliffs on either side.

What a terrific country this must have been not so long ago as geologists reckon time: ten, perhaps twenty thousand years back! I began dimly to perceive what it must have looked like before these deep valleys were gouged out by the rivers. It was a plateau with a general elevation of about 12,000 feet, but with a steady slope to the south, crossed by ranges of mountains, some of whose peaks rose as high again. On the south side, glaciers as much as seventy miles long crept down towards the plains; and the rivers to which their melting gave rise, cut out the lower valleys such as the Lohit, and laid down thousands of feet of silt, layer by layer, which to-day make up the plain of Assam. There was not then such a network of ranges as we see to-day, because there were not so many streams. A thick skin of ice covered the country, filling the hollows. To-day, the lesser ranges *are* ranges: but before these valleys were cut, when the general level of the country stood higher, the lesser ranges were no more than ridges, buttressing a single great main range.

There was as yet no obvious change in the vegetation as a whole, that is to say, we were still in a land of forest—Pine forest down in the valley and on exposed slopes; temperate mixed forest on the higher ranges, where the snow lingered and the air was cooler and also moister. Under the Pines grew masses of *Stellera Chamaejasme* and clumps of dwarf purple Iris, pink Martagon Lily, and yellow violets.

There were low stone walls round the fields to keep out the cattle. Masses of white-flowered Jasmine and the grotesque and foetid *Aristolochia Griffithii* sprawled over these walls, and in the woods grew a number of ground Orchids, with small but dainty flowers. Perhaps the most interesting forest plant, to a botanist at least, was a perennial herb with large pale green compound leaves and panicles of rather inconspicuous yellow flowers, followed by a pyramid of conspicuous bluish fruits, hard and spherical as stone marbles. This plant, *Caulophyllum robustum,* is allied to the Barberries. It was by no means common, and having found a few plants, which I gathered for pressing, I found no more till November, when I recognized some fruiting specimens.

One sunny day we went down the valley to take pictures of the gorge and of some shimmering cascades. Having descended to the bridge, we scrambled down to the water's edge, and reached, with some difficulty, on account of giant stinging nettles and thorns, a commanding position from which to photograph the bridge. Returning, we decided to evade the vegetable armament by wading under a cliff, where there was slack water. The water was crystal clear, the sandy bottom further away than it looked, and I easily stepped into a hole waist deep. But it was not the depth which shocked us: it was the temperature. Never before have I waded through water so icy cold that, even when immersed no more than knee-deep, I instinctively caught my breath. It literally made me gasp, and though I sat in the sun immediately we got out, it was ten minutes before the numb pain in my legs disappeared.

Ata is a miserable, muddy village of tumble-down houses, but there is a good cantilever bridge over the river, and the road to the Ata Kang La continues up the left bank. The water was grey with glacier mud, and rising daily. It ran like a mill-race.

The day after we arrived at Ata I walked up the valley to look at the glacier. A wall of snow peaks filled the narrow horizon, and the glacier wound upwards till it was lost to view round a cliff. When I reached the foot of the glacier, I found another large valley, also filled by a glacier, which flowed from the north-west: the two glaciers did not quite meet, but the milky torrents to which they gave origin united where vast mounds of barren moraine were piled up. Ascending the moraine, I looked up the western valley, and for the first time caught sight of a magnificent rounded snow peak from which one of the glaciers flowed. The peak is called Chömbö. The Tibetans say that it is the highest peak on the range, a statement which must be accepted with caution. I estimated its height at not less than 22,000 feet. A cattle path led up the north-west valley, skirting the glacier; I followed it, ascending steeply through the forest. There were glorious old trees here, their twisted and gnarled branches interlocking to form a high cathedral-like roof: Juniper, Maple, Spruce (Picea), Rhododendron and Tsuga appeared: then Larch, *Gamblea ciliata, Magnolia globosa,* and Birch, all trees of great size. Beneath the trees, the spongy ground bore quaint Arums (Arisaema), tall pagoda-like lilies (*L. giganteum*), *Podophyllum versipelle* with peals of bell-shaped white flowers hanging from under flat umbrella-shaped leaves, and ferns. By the interlacing streams which gurgled through this forest of Arden, the coppery-orange flowers of *Primula chungensis* were impaled, whorl by whorl, on mealy stems. Suddenly I emerged into an open meadow; the forest shrank back. Yak were lying down on the snow which spread out fanwise into the meadow from a gully; it was their way of keeping cool and a protection against the myriad sandflies which clouded the damp atmosphere. I had found a likely camping ground, from which the alpine region might be reached.

So far I had been walking along the top of an ancient lateral moraine, high above the glacier. I followed the bank for a short distance beyond the meadow, still buried in luxuriant forest, wading knee-deep through the lush undergrowth. Then the moraine came to an abrupt end. Through a gap in the trees I looked straight down on to the broad glacier 200 feet below; and up the valley to a magnificent view of Chömbö, framed between wooden pillars. To ascend the valley further, it would be necessary to scramble down on to the glacier, and toil slowly up its rough surface; for the steep cliffs were now flush with the ice, and the lateral moraine disappeared between the glacier and the rock wall.

Satisfied with what I had seen, I returned to Ata. I had asked Ronald to go up the main glacier valley and report on a possible camping place, which he did. But on comparing notes, I preferred my meadow, and laid plans accordingly.

It was now the first week of June. The Tibetans told me I could not possibly cross the Ata Kang La till July: we had therefore four or five weeks in which to finish the picture, and botanize in Zayul. After that, I would continue my journey to Shugden Gompa, while my companions returned to Rima by the way we had come, and thence to India.

At Ata we found an aged craftsman skilled in turning those wooden bowls, called by Europeans *tsamba* bowls. Every Tibetan, however poor, possesses a *tsamba* bowl. But they vary in quality. Some are made of walnut, and some of commoner wood. Those of the well-to-do, of officials and high priests, are lined with silver. The very best are made of a species of maple (*Acer stachyophyllum*), which grows scattered in the temperate forests between 8000 and 10,000 feet altitude. It was not uncommon above Ata. This maple is a remarkably handsome tree, not very tall, but with spreading branches. Externally the most striking feature is the apple-green bark finely streaked with mutton-fat white, as though a milky latex had exuded and trickled down the outside: this stippled bark conceals a no less beautiful grain in the wood. The leaves are polished, bottle-green, and simple, the flowers hang down in long racemes. When the half-ripe fruits appear in June, the tree is particularly handsome; for the fruits are

THE WOOD TURNERS

smouldering red, and the branches hung all over with these long dully-glowing tassels are a memorable sight. The craftsman, with his crude lathe, was not less wonderful than the maple wood he turned. The lathe was made entirely of wood, and the object to be turned was attached to the chuck—if it could be called such—with a black gum, softened by heat. It was a reciprocating lathe, worked by a treadle: a small boy leaning on a bar did the work, while the craftsman, seated on the ground, with six long bayonet-like tools of soft iron arranged beside him, made the shavings fly. He held the tool under his armpit, pressed firmly against the wood as it spun to and fro. He could turn out a cup well-shaped, smoothed and polished very quickly. The varnishing was done elsewhere; a high polish is put on. It is a quaint reflection on our common human nature that the Tibetans are beginning to buy enamelled iron *tsamba* bowls, made in Japan, apparently preferring them to these lovely wooden bowls. You cannot even hold an enamelled iron bowl filled with scalding tea, much less drink out of it. I heard a strange story in Ata. It is said that in winter large birds fly south from the plateau, over the Ata Kang La. No sooner do they reach the snow, however, than they come down and *walk* over the pass. When they have crossed the snow pass, then they take to the air again, and fly down the valley by day. From the description given of these birds, they are probably pheasants. The only pheasants I saw here were monals, and monals certainly do walk rather than fly at high altitudes. But I cannot vouch for the truth of the story.

Although Ata lies in a deep glen, at a height of less than 8000 feet, it was quite cool even in June. All round us the high peaks were covered with winter snow and a cold stinging rain swept remorselessly over the Assam hills. The rain loosened the snow, which roared down the high gullies, kicking up clouds of frozen spray which hung stiffly in the air. The glacier which filled the upper valley from wall to wall also absorbed a vast amount of heat; and beyond its sources rose the everlasting snows of Tibet. Far down the valley towards Rima spring had been left behind; it was already summer. But the advance of summer into Tibet is a slow business, and we had outstripped it; we were once more ahead of the clock. However, trees and shrubs now began to surge into bloom, and the dark forest was spouting fountains of flower from the Indian Ash (*Fraxinus floribunda*). The village itself was embowered in pink roses and a fragrant ivory lather of Mock Orange.

THE RIVER OF ICE

Meanwhile the villagers had gone up to the meadow I had found, to build us a small shack. As soon as they returned we moved up into camp. That was on June 8th; after a lot of rain, the weather showed signs of clearing.

CHAPTER VII

PLANT HUNTING

JUNE 9TH was a beautiful day. I decided to climb the snow gully which watered our meadow, this being the most direct route to the alpine region. It was extraordinarily steep, but not otherwise difficult to begin with; the snow was soft, so that I could kick steps in it as I ascended. Coming down, I thought, would be easy—perhaps too easy. Presently matters became more difficult. The slope grew steeper, the snow more compact and slippery. Looking down the gully, I began to have qualms. If I slipped, I would shoot down the slope, gathering speed as I went. Moreover the gully was not straight; it twisted, and if I slid down out of control, I would probably bump up against the cliff at the first turn, or what was just as bad, disappear into the deep crevasse between the snow and the rock wall. I did not much like it, but the walls of the gully were almost sheer, and I saw no place where I could climb up. So I continued up the snow cone, rapidly approaching its apex as the gully narrowed and steepened. In this wise I ascended a thousand feet. Eventually I found a place where I could step off the snow cone and scramble up the steep wall on the sheltered side, which though almost sheer, was covered with scrub: the opposite wall, being exposed, was naked. The scrub consisted chiefly of Rhododendron; and Rhododendron scrub, growing on a precipitous rock face, forms quite one of the most impenetrable barriers in the vegetable kingdom. I had a hard tussle to get through. Here grew *Rhododendron cerasinum* (crimson-scarlet flowers), *R. cinnabarinum* (flame flowers), *R. pruniflorum* (plum-purple flowers) and *R. trichocladum*, a bare shrub, flecked with pale yellow flowers before the leaf buds had burst. In a cleanly scoured couloir, from which the snow had disappeared, clumps of pigmy violet Iris encrusted the rocks, and a small form of *Primula sikkimensis* was striving to flower. At last I was on the threshold of the alpine region. Returning to camp, I found that a pig, ordered from the village, had arrived. We built a large fire in our shack, which we used as a mess room, ordered a big dinner, and spent a pleasant evening. It was a tranquil night, the stars radiant overhead: a cold wind off the snow blew through the shack, but we kept some of it off by hanging up a couple of waterproof canvas sheets. At ten o'clock we retired to our tents, which we found infested with earwigs.

Two days later I made another successful attack on the alpine region by the snow cone. This time I climbed out of the gully lower down, and ascended a forested buttress. The buttress was precipitous, but being forest-clad was nowhere difficult; where there is forest there is little undergrowth. There were many big trees, and *Rhododendron fulvoides* was still in bloom. When I reached the treeless slopes again I found them so thickly clothed with Rhododendron tanglewood that there was standing room only.

R. cinnabarinum, in every shade of salmon and apricot, was a gorgeous sight. I pushed through to a gully, now dry, where water, and snow, and falling rocks, kept a passage open. It was lined with stunted bushes of *R. Beesianum*, each bobbing head of pink flowers surrounded by a collar of long downward pointing leaves. At an altitude of 13,000 feet I found deep snow, but also beautiful dwarf Rhododendrons in bloom: the crouching *R. sanguineum*, incandescent with blood-red flowers, and aromatic brooms of *R. kongboense*, its coral-red flowers clustered into orbs, which shook in the wind. I had now been climbing for over four hours, and was very thirsty, for the sun beat fiercely down. There was no water, the snow simply evaporating. I observed that Chömpö had twin peaks, one south, which we had already observed, and one behind it to the north; but it was impossible to say which was the higher. From the south peak the smaller glacier rose; the big Ata glacier had no connection with Chömpö, which however must give rise to a large glacier on its north face. I returned to camp by the way I had come.

Meanwhile I had sent Tsumbi up the main valley to a camping ground called Chutong, on the road to Shugden Gompa. I bade him report on the snow, and whether we could pitch a camp there or not—the local inhabitants having said that this was impossible. He returned with a favourable report. Chutong, he said, was just on the tree line about 13,000 feet. There was a great deal of snow yet, and he had been quite unable to go any higher, but he thought it would be possible to build a shack for us. This sounded good. We returned to Ata: and I told the headman to take a working party and prepare a camping place.

Four men departed with axes and food; and we settled down again to our work.

By June 15th the flour we had brought with us from Sadiya in sealed kerosene tins was finished, although lately we had eked it out by mixing it with ground rice. It had been packed damp, and some of it had gone mouldy; not that we minded much. We now fell back on *tsamba,* pressed into cakes; it was good, nutritive stuff, but not very palatable. Later we got some coarse flour from Modung. One of the lamas resident in the village brought us some delicious meat dumplings, made of chopped mutton and red pepper wrapped in soft hot pastry.

On the edge of a cultivated field just above the village I had found a colony of *Lilium Wardii* in bud—a dwarf and probably starved form, like that growing in the Pine forest above Modung; though this was the only occasion on which I ever saw *L. Wardii* growing right out in the open, on flat ground. It was now in bud, but did not look like flowering for some time. I therefore decided to go right down to the river gorge below Modung and collect it there, thinking it would certainly be in flower by now; three weeks ago, fat buds had promised to open shortly. So Tsumbi and I rode over to Modung and started down the gorge on foot. But it took us a long time to reach the cliff where the pink Martagon Lily grew, and when we got there it wasn't in flower. It was nearly seven o'clock when we got back to Modung. We halted for a meal and a rest. The headman urged me to stay the night, but I was determined to get back to a comfortable bed, and declined. By the time we reached Gechi Gompa I almost regretted my obstinacy. It was dark. Rain had begun to fall and the little monastery was deserted. For the next mile the rough path along by the river proved very difficult, and we could only move a step at a time. Under the trees it was dark with an opaque and menacing blackness. I felt as though I wanted to cry out in sheer desperation. We stumbled stupidly along in single file, the glacier river booming along close beside us. One false step over the edge and this terrifying uncertainty at any rate would be over. Suddenly I saw a light in a hut up the hill, and sent Tsumbi to investigate. He returned with a pine torch and a half-wit who knew the way. We crawled on, wet and weary: it was easier now with the torch. At last we came out of the forest on to the open path. A pale mist hung low over the river clearly defining it. Soon we reached the bridge, and a few minutes later we were back in the village. It was nearly eleven o'clock, and my companions had wisely gone to bed, supposing me to be spending the night at Modung, as I ought to have done. Late as it was, Tsumbi managed to get drunk and make a nuisance of himself.

On June 18th the working party returned from Chutong, having built a hut for us; and I gave orders that we would proceed with all hands on June 20th, leaving behind only such stores as Ronald and Brooks-Carrington would require for the return journey to Rima. Luckily June 20th turned out an exceptionally fine day; and after taking some pictures in the village we started up the left bank of the valley, soon finding ourselves on an old moraine high above the glacier, and deep in the forest. Occasionally, from the edge of the moraine, we caught a glimpse through the trees of the glacier winding mile after mile up the valley. Looking downstream, I observed a curious thing. On the right bank, towards the snout of the glacier, was a moraine of reddish earth and rock which formed an arc, curving in towards the centre of the glacier. A similar arc, composed of greyish moraine, had formed on the left bank; the two arcs met at the snout of the glacier. But this was not all. There were in fact three concentric moraine arcs, produced by the convergence of lateral moraines, which then combined to form a sort of terminal moraine, although the ice continued to flow beneath the lowest arc. Each of the two lateral moraines split into three near the snout as the glacier swung in from the cliffs; but each of the three stood on a different layer of ice, and at a different level. The glacier during its intermittent retreat had left terraces, exactly like a river; and these concentric arcs were more of the nature of terraces than moraines, marking stages in the retreat of the ice.

There were two well-defined derelict lateral moraines on the side we were ascending, both covered with dense forest. Many of the Abies and Picea trees were fully two hundred years old, which gave some evidence of the age of these moraines, now over a hundred feet above the glacier.

The first trees and shrubs to establish themselves on the abandoned moraines are Poplar, Buckthorn, Tamarisk, and a species of Honeysuckle. Small moraines, covered with a dense scrub of Buckthorn, extended for half a mile down either side of the main glacier, below its snout, towards Ata; but as the glacier retreats, the material is rapidly washed away, and re-sorted. Only exceptionally is a moraine preserved in this climate, as below Rima. On the moraine at the ice foot, in the barest spot, I flushed a nightjar off its nest—a mere hollow amongst the stones containing two eggs. We camped the first night in a small clearing in the forest called Shukdam. Next day the real excitement of the climb began. We soon left the brink of the glacier and ascended the great south face of the range, which grew rapidly steeper and rockier. At last we emerged on to an open slope which, though well below the tree line, was covered with alpine flowers. A species of garlic grew here, and we all collected it to season our food. From this point we had a fine view up the glacier, with a needle-shaped peak beyond. Sheer black cliffs, striped with snow, hemmed it in. After resting and eating our lunch on this ledge, we resumed the climb up a tall scree. There was a huge devastated area here as though giants had been demolishing castles up above. Millions of tons of rock had fallen, and up this washout we climbed slowly. Then traversing through a belt of Rhododendron and Silver Fir, still following a rough path, we climbed a broken face, where yellow Rhododendron bushes were flowering gaily, and reached our hut, on the last fringe of trees. We were hot after the long waterless climb, and it was not till the sun had set behind Chömpö that we realized how cold it really was. The altitude of Chutong camp was 13,129 feet, and deep snow lay all around. Close beside the hut was a piece of ground free of rocks and shrubs. We levelled it as best we could, and pitched our tents here, using the hut as a communal apartment. A second smaller hut alongside was used as a kitchen and sleeping place for the men. It was our first permanent camp, and we made ourselves as comfortable as circumstances permitted.

Next day most of the coolies went down to the valley. Two we retained, one to cut firewood, the other to bring water. As a matter of fact there was no water; our source of supply was the snow, which we melted in buckets; anyhow there was plenty of it. Three men who had brought guns and hunting dogs stayed on for two days, but though they got on the track of a gooral, the dogs lost it. The next day, June 22nd, was our last fine day for a fortnight. Photography was now relegated to the background, and I devoted my time to plant hunting; but there was so much snow that even here I was handicapped. However, time would cure that; the snow was melting fast. The last of the forest trees were Silver Fir (*Abies Webbiana*) and Larch (*Larix Griffithii*), with an undergrowth of the big leafed *Rhododendron Beesianum*. In the open, amongst the chaotic boulders, *R. Wardii* formed domes of sulphur,

MY COOLIES ABOVE THE ATA GLACIER

flecked with coral red buds. At higher levels heaving seas of dwarf Rhododendron were mixed up with the snow; rose-pink, crimson and purple flowers floated and bobbed on the crests of the waves.

By the end of June there were at least twelve species of Rhododendron in bloom in the alpine region, ranging in colour from deep Tyrian purple, plum-purple, cherry-red and scarlet to the most delicate apple-blossom pink. Even more attractive were the alpine herbaceous plants. Amongst the boulders, a Cambridge blue biennial Poppy (*Meconopsis horridula*) was fluttering its petals; and a mop-headed Primula, clouded with white meal which showed up the mauve flowers, grew on the screes. This is a form of the well-known *P. denticulata*. Shaggy, white-haired Anemones covered the alpine turf slopes in thousands: the white satin flowers have a nacreous lustre, like moonstones. Mixed with them were shiny-leafed plants of *Nomocharis Souliei*, each plant bearing one Lily-like flower of a deep port wine red colour; an interesting plant, but not at all lovely. On these steep south-facing slopes the snow had melted and a variety of meadow flowers were coming up. Last year's dead vegetation still lay prone, not much altered. The soil beneath was pure humus, derived from the slow disintegration of the vegetable debris. The weight of snow presses it into a solid mat and doubly preserves it, preventing decomposition, as well as packing it. In spite of a liberal mixture of angular stones it retains plenty of moisture and eventually forms an unctuous black soil, highly charged with organic matter. Chutong attracts a very heavy snowfall, but most of the snow melts again. One of the first things we noticed here was the measureless silence of the mountains. For the first time for weeks we were beyond earshot of the roaring torrents. All movement, save the sudden brief movement of the avalanche, was clamped by the cold, which held all things in its vice-like grip. But great disturbance was going on beneath the earth's surface. The snow was fast melting: millions of seeds were germinating, millions of tiny plants swelling and pushing upwards through the softened soil. And so steep was the mountain that as the snow became loosened beneath by the restless vegetation, and detached at the rocky sides by melting, it slid off in powdery masses which roared down the precipices. Then silence again.

The rain came. Billows of cloud swept out of the west and hung like wet smoke over a flabby world. Early in the morning the mist rose out of the fathomless depths below where it had laid all night like cotton wool. It smothered the forest and left it dripping. It formed and took visible shape out of the air and hid the snow peaks and glaciers behind a veil. And always the rain descended from it, not heavily like a tropical storm, but steadily, everlastingly. Our camp was pitched two thousand feet below the crest line of the great ridge, beyond which was another deep valley also occupied by a glacier. This glacier flowed from the Ata Kang Pass and the high peaks to the south-east of Chömpö and joined the main glacier far below us, out of sight. The only possible route down to the main glacier was the way we had come up; above the point where we had left it, it flowed in an impassable gorge. The only route down to the tributary glacier was over the ridge, and down the other side on a long traverse to its upper course; for it, too, plunged into a gorge below.

Ronald and I had not been a day at Chutong camp before we had climbed the ridge, tramping up through the deep snow to the notch called the Cheti La. The view from there convinced us that it would be a fortnight before we could cross even the Cheti La, let alone the main pass which loomed ahead through the mist. The range bristled with snow peaks, short glaciers hung from the black cliffs. The descent from the Cheti La was precipitous for some hundreds of feet—a chimney, packed with hard snow. Nor was there any sign of animal life in this frozen wilderness, except a family of monals and a pair of snow pigeons (*Columba leuconota*). But as summer advanced, bringing the warm rain from India, it was very different at our camp. More and more birds flitted amongst the sodden Fir trees—rose finches, yellow breasted robins, flycatchers and speckled babblers. Small birds constantly visited the Rhododendron flowers, burying their heads in them to get at the insects which are to be found feeding on the honey, and would come out with their heads dusted with pollen and go off to another flower, thereby pollinating it. *Rhododendron Wardii* and *R. Beesianum* were constantly visited by birds. The soft earth too was being churned up by voles, which occasionally showed themselves by day, while pigmy hares commonly did so. They, however, sport on the boulder heaps, ready to dart swiftly into a crevice on the first hint of danger—though being incorrigibly curious, they quickly come out again. Sometimes we heard the cuckoo, or the sharp minatory cry of a pheasant.

A favourite climb of mine was to the summit of a high cliff which overlooked the glacier junction. Here a seam of limestone is embedded between the slates and schists which flank the granite core of the range. It crops out like a groin. On the crisp turf brilliant alpine flowers grew, notably a fragrant mauve flowered dwarf Primula, and the refined *P. sikkimensis,* Globe flower and purple Orchis. On the cliffs, tufts of the incomparable *Paraquilegia microphylla* trembled with every passing breeze, and plumes of fierce orange flowered Oxytropis challenged the climber. The view of Chömpö and a whole arc of the main range east of the glacier pass was terrific. I enjoyed standing up there on the topmost pinnacle, 3000 feet above the tempestuous river of ice, watching the drama of the storm clouds as they smote against the peaks which rose 7000 feet above me. How high they rose above the petty ways of earthbound man! One day the Tibetans brought up a pig I had ordered and a few eggs at the same time. We still lived chiefly on rice, raisins and currants—which make an excellent curry, with *tsamba* cakes; we had taken to flavouring the *tsamba* with cinnamon, the result being known as cinnaba. The only local additions to the larder were garlic, of which there was an abundant supply, a few bamboo shoots from the forest, and an orange fungus which grew on the Fir trees. This last, boiled and cut into strips, though rather leathery, had a pleasant flavour.

On July 5th I sent Tsumbi down to gather flowers of *Lilium Wardii* and to bring up the coolies. It was time to make a further advance towards the Ata Kang pass. After two false hopes, the 8th turned out a really fine day with brilliant sunshine. Brooks-Carrington and I made the most of our opportunity, and took some hundreds of feet of film.

CHAPTER VIII

FILMING THE FLOWERS

Since the day after our arrival at Chutong camp, when Ronald and I had visited the Cheti La and found it impassable, I had climbed up several times at intervals of a few days to see how fast the snow was disappearing. On June 28th we crossed the ridge and descended the chimney a little distance; but on account of an overhanging snow cornice, we had to clamber up and down the cliffs, a hazardous undertaking impossible for coolies carrying loads. Like the dove from the ark, I went up again on July 1st to see if the snow had melted; it had not. On this occasion I saw three female monals with four chicks just able to fly. They walked up the scree, uttering their plaintive cry.

During this fortnight in the clouds we adopted a regular routine. We rose between 6 a.m. and 6.30, and had a cup of tea in the shack at 7. Then we worked at anything we happened to have on hand till 9, when we breakfasted. Between 10 and 3 I was usually out climbing, searching the cliffs for new plants. At 4 we had tea, and then we usually took things easily for an hour, perhaps reading a little. Dinner was at seven; and after talking or playing a game of chess, reading, and writing our diaries, we turned in. Brooks-Carrington devoted a lot of time to his camera, or to making some ingenious contrivance, or mending our gear. He was a first-rate craftsman. Before the servants laid our morning tea, they chanted long prayers and lit a pile of Juniper branches outside which volleyed smoke into the flabby air.

After our fine day on July 8th the weather returned to normal. Tsumbi got back that night. Instead of collecting the Lily at Ata, as I had told him to do (describing the patch), he had gone right down to the gorge below Modung for it. Anyhow, here it was, undoubtedly a form of *Lilium Wardii*, deliciously fragrant. It surprised me to find so many scented flowers and leaves here; to mention others, there were two species of Nomocharis, *Primula sikkimensis* and another Primula, a scrub Honeysuckle with small pink Daphne-sweet flowers, and several aromatic Rhododendrons. As for the Nomocharis, a word may be said about these. They are Lily-like, alpine, or occasionally sub-alpine bulbous plants, peculiar to these mountains. They are most abundant in regions of great precipitation, especially where snow embalms the alpine vegetation for months at a time. I found three species at Chutong. The commonest was *Nomocharis Souliei*, already referred to. Much rarer was a fragrant yellow flowered species with grass-like leaves. It was abundant on one particular scrub-clad slope, but I did not see it anywhere else. Probably it was a form of the widely spread *N. nana*. A third still rarer species also had fragrant flowers, but they were plum purple, and the leaves were shorter and broader than those of the last. All three plants bore no more than one, or occasionally two flowers, whereas *N. pardanthina*, which I found growing farther south, above Solé, in December, bears six or eight flowers, and in fruit looks much more like a Lily.

Before Tsumbi had gone down the valley to gather *Lilium Wardii*, I had warned him that he would probably have to return with my companions: I could not, while alone, be saddled with the responsibility of a man who got drunk. Tsumbi, of course, was my No. 1 servant, and it was only right that he should stay with me when the others went back. Only under exceptional circumstances would I have questioned that right. But on three occasions Tsumbi had been drunk, and when he was drunk there was no disguising the fact. Far from hiding himself, he came out into the open and threatened war on all and sundry, with tiresome garrulousness. Since warning him that I would probably take Pinzo and send him back, Tsumbi had been very zealous: and when he returned so quickly with everything done, I told him that I had decided he should accompany me to Shugden Gompa. The real truth was, I had to send a reliable man back with my companions, and Pinzo was the best all round man for so rough a journey.

On the 9th the coolies arrived, bringing us a fresh supply of butter, two fowls and eighteen eggs.

On the 10th we set out to cross the Cheti La. It was a dull day, but not actually raining. After descending the chimney, now clear of snow, we traversed along the base of the cliffs, keeping high above the glacier. Scarlet Runner, as Cawdor and I used to call *Rhododendron repens*, was abundant here, a tiny creeping plant with enormous scarlet trumpets which lie flat on the cold stones, always pointing down the slope. After ascending a moraine we climbed down to the glacier and pitched our tents on a level shelf covered with stones. There was no firewood here, nor indeed any vegetation at all. Everything was under ice and snow: but we had brought up enough firewood with us for the night's cooking. It was a dull evening, a cold wind blew from the pass, and our tents looked very uninviting. However there was nothing for it, and after a tepid supper we turned in. The next day I sent six coolies down to the forest to bring up firewood, and announced to the headman that we would stay here several days; they could return to Chutong and remain there till I was ready to cross the pass, getting up rations from Ata as required.

I could have crossed the pass now; in fact a party of men had crossed as early as July 9th, the day before we started from Chutong. But I tarried here on the cold glacier for five days, in order to get colour pictures of the flowers, as well as of the magnificent scenery. At first it looked as though we should never get any pictures at all. The weather was vile. But Brooks-Carrington and I persevered. There were three flowers I was especially anxious to film: the glorious *Paraquilegia microphylla*, unsurpassed as a cliff plant, Scarlet Runner, hissing in red rivers down the wet screes like molten lava, and *Rhododendron sanguineum,* so scalding red hot that it seemed to clear a space for itself by melting the snow all round. These three vivid plants grew down the valley, where the path traversed at the base of the granite cliffs, and on the way up I had selected plants to photograph. On the 13th, after a night of terrific rain, we returned to the site. We had just set up the camera and taken one shot when, almost without warning, a dreadful storm off the snow peaks assailed us. A tearing jagged edged wind swept over the glacier; in a moment we were enveloped in freezing cloud. Hastily covering the camera with waterproof sheets, we turned our backs to the storm, and sat huddled and miserable on the steep slope. The rain sluiced everything and battered our bodies to a numb pulp. Only our hearts held on. So we sat for two hours. In the late afternoon we gave it up. We reached camp gelid with cold, and drank hot tea and rum in our tents while we changed our sodden clothing; it required hot stones to restore my circulation. Even as we sat there, enjoying the comfort of our tents, the storm passed. A cold draught of air came over the pass from the north; one by one the great peaks showed up, the clouds lifted and were stretched out in long level bands, a few stars shone bleakly. An avalanche roared down, splitting the silence like a wedge. We put on all our warm clothing, and dined comfortably; then crept into our sleeping bags, happy at the thought of a fine day on the morrow.

It was a beautiful dawn. While we were having breakfast there came an ominous crack high up on the cliffs above our camp. We dashed out of the tent and stood spell-bound watching thousands of tons of snow and rock sliding down a gully opposite to us. Huge blocks of stone muffled in a lather of powdery snow gathered momentum and rushed down on us. Only the ridge of the moraine intervened. It was sufficient. The mass turned aside and gradually came to a standstill.

Brooks-Carrington and I returned to our vigil. The sun shone intermittently, and we took several shots. But there was an ugly look about the sky. The shot we had tried to get the previous day, a cascade of scarlet Rhododendron blossom, was left till the last. We had just set up the camera at the old spot once more when the sun disappeared. Dark mists came rolling like smoke up the glacier and a tongue of forked lightning zipped out of the formless void. Instantly came the crash of thunder, followed by terrific rain. Again we cowered against the mountain side, getting slowly wetter and colder and still wetter, till the water ran out of us. We hung on, but the storm grew fiercer, and there was never a hope of getting that particular shot. It was dusk when we got back to camp. The storm was over. A tempestuous calm succeeded. The peaks were reappearing, dead white in their clean new shrouds: they looked ghastly and corpse-like. Suddenly the tip of Chömbö itself appeared, floating on a bank of cloud, the rest of the mountain being hidden. It was a wonderful vision, just that dead white cone, crystallizing out of the shapeless mist, faint at first then more sharply outlined. I took it for a sign; a promise for the morrow. And so it proved.

It was difficult to believe that the altitude of Glacier Camp was barely 14,000 feet, so frigid was the scene. Just above us the glacier slid down a steep slope, and the ice became broken up into *séracs*. Every hour or so, alongside us, a *sérac* toppled over with a roar. The glacier, too, creaked and groaned in the night; it was much noisier here than at Chutong. More exciting to watch were the avalanches. There were three hanging glaciers visible on our side of the valley, jammed high up into gullies. From time to time the end of a glacier split off, and came crashing down; or a higher gully tipped its snow into a lower one, as a grocer pours castor sugar into a tin. Anything more repellant than the

Ata Kang La it would be difficult to imagine; repellant and dangerous. No wonder the pass is only open during four months of the year and is little used. Few people are lost on it. We could not quite see the summit from our camp, owing to the ice fall which rose immediately above us; but I had seen it from the Cheti La. Looked at from there, and still more from our present camp, it seemed a most unlikely route. Glacier Camp itself seemed to be the Ultima Thule. Above us there were no more lateral moraines; it was up the glacier and over the ice fall, or at any rate across the glacier to the cliffs on the far side, where there was a rough track part of the way. But once above the ice fall, the glacier, though crevassed, certainly presented no difficulty. During these last days my thoughts continually strayed to the mysterious land behind the snow peaks; in spite of the storms which were our daily portion on this side, and the great banks of cloud which rolled up the valleys, the sky over the pass was always a serene turquoise blue. There at least the sun shone day after day; the clouds seemed unable to cross the barrier.

In that fact lay the answer to much that had hitherto been obscure. Zayul, to the south of the range, is a land of forests. It is inhabited by people who live by agriculture and by hunting. On account of the heavy snowfall, cultivation cannot be carried on at a high altitude, and as a matter of fact, there are few villages above 7000 feet, and none above 9000 feet. The deep valleys are very hot in summer; hence rice is the principal crop up to 6000 feet; above that, maize.

North of the range is a very different country, a pastoral land, without forests. The climate is severe, there is little snow to protect the vegetation, or to keep it moist in the early summer. The average altitude is between 12,000 and 14,000 feet, and even in the main river gorges there are no villages below 10,000 feet; on the plateau the villages stood above 13,000 feet. Barley is the only crop.

That I stood just below the crest line of a great mountain range was obvious. There was at least one peak, Chömbö, not less than 22,000 feet high; but I had seen a dozen others of the order of 20,000 feet. The great Ata glacier which rose to the north of Chömbö I estimated to be 12 miles long, and its average width I found by measurement to be 900 yards. It is the largest known glacier east of the Tsangpo gorge. Two other glaciers were the Chömbö glacier, about 8 miles, and the Ata Kang glacier, about 6 miles long. These figures are startling. No such large glaciers had ever been suspected so far east; although, as I have shown, they are mere remnants. Evidently then this range is an important one. The question I now wanted to solve was, what became of it to the north-west?—and also to the south-east? As to the formation, the core of the range, and the high peaks, were of gneiss and granitic rock with dark fine grained pegmatite inclusions. Lower down the valley, towards Ata, that is on the flank of the range, were altered sedimentary rocks which dipped south-west at varying angles. Even at Chutong limestone and slate cropped out. At Modung were mica schists, as well as granite.

That night, July 14th, we had our last dinner together, and I gave my companions full instructions how to get from Rima to Fort Hertz, in Upper Burma. I had myself done the journey three times within eight months, twice in the rainy season and once in the cold weather; so I was able to tell them my own experiences. It is not a difficult journey, though it is apt to be troublesome at certain seasons. However I had no qualms. We had a last talk, and then went to bed. I had given instructions for the coolies to come early next day, as I was now ready to start for Shugden Gompa.

CHAPTER IX

OVER THE GREAT SNOW RANGE

NEXT morning we rose at six to find the mountains hidden by mist. While I was dressing, rain began to fall, at first in a drizzle, then heavily.

At 8 o'clock, after we had finished breakfast, the rain stopped and the clouds began to clear so fast that the sun soon blazed out in a hard blue sky.

My clothes were sopping; so I decided to wait for Brooks-Carrington to take some pictures before I started. While my clothes dried he took shots of the camp. When we set out, Ronald wished to come as far as the pass, the frontier of Zayul, beyond which I had promised the Governor of Zayul I would not take him. Brooks-Carrington came half a mile up the glacier in order that he might take a last shot of the party struggling up the ice-fall.

I said good-bye to Pinzo with some regret: he was one of the best servants I have had. I was depressed, too, at the thought of being alone for such a long time without the company of white men. My party consisted of nine coolies, Tsumbi and Tashi.

We set out at 11.15, keeping to the centre of the glacier, on which the snow was firm and our feet did not sink in deeply. Except for one place where we had to ascend amongst séracs the slope was easy; but as we approached the pass, the glare from the snow became dazzling. I had snow-glasses, and had provided a pair each for Tsumbi and Tashi. The coolies were not so lucky, shielding their eyes by pulling their hair and hats down, some even tying a flap of cloth over their foreheads.

The top of the pass I estimated at 16,000 feet,[1] and we took two hours to reach it. The pass is a broad saddle between high rocky peaks surrounded by glaciers. The Ata Kang glacier rises from a peak just to the south-east and flows steeply on to the saddle, whence it radiates more gently down three separate valleys. One of these is the one we had come up from the Cheti La, so that its water eventually reaches the Ata river. A second and shorter branch flows north and is one of the sources of the Nagong river; a third flows westwards, but there is no route down it.

Having said good-bye to Ronald, who was as grieved at having to go back as I was to lose him, we started to descend. First came snow, which was easy work. But we soon got on to slushy ice, sprinkled with stones and fine gravel, down which a multitude of small streams flowed. Here we sank in over our ankles.

After a couple of miles or less, the glacier ended in a piled mass of rock and stones, stretching a distance of over two miles; through this moraine, streams from the glaciers have cut deepish courses, uniting lower down in a milk-white torrent. There was no path defined, and we had to scramble over the rocks as best we could: slow and heavy work for the coolies with their loads.

Plants grew right up to the moraine, just scattered tufts of grass at first and scrawny shrubs, willow and tamarisk.[2] But we soon came to richer vegetation and more varied. Rhododendrons of course; the purple flowered *R. telmateium,* and bushes of *R. sigillatum,* stiff leaved, cowering in holes among rocks; *R. crebreflorum* and *R. riparium,* with flowers of brilliant magenta. We soon trod firm turf, on which were flowering squat vetch-like Oxytropis, the colour of burgundy, and yolk-yellow Astragalus. Then bright pink *Androsace Chamaejasme,* and yellow Potentilla and Ranunculus.

It was surprising how quickly we passed from bare rocks to a varied alpine flora. A mile lower still the variety was even greater; Asters, Anemones, Primulas, Polygonum, Trollius and Cremanthodium being among the most notable. The banks of all the many streams which entered and flowed down the valley were marked with lines of flowers.

Coming to where the slopes were broken and covered with shrubs we halted. The torrent here ran swiftly between low level banks; and the valley broadened to half a mile in width. Yak were grazing right up to the foot of the moraine. We camped by a side stream overhung with willow bushes beneath which *Primula szechuanica* grew in profusion. This Primula sends up a two-foot stem, which bears one or two whorls of butter-yellow flowers, which would be large but that the petals are reflexed. Their scent is very fragrant. The leaves are malachite-green and strap-shaped, silvered beneath and like wash-leather to touch. Here too was *Primula cyanantha,* with downy leaves and murky violet flowers in club-heads, more like a grape hyacinth to look at than a Primula.

On the slopes grew thickets of *Potentilla fruticosa*, bearing white, cream or bright yellow flowers; a Spiraea whose curving branches were thick with blobs of dusky red bloom; *Caragana jubata* bristling with thorns, spiny Berberis, Lonicera, Juniper, and other shrubs.

But I was not interested so much in these shrubs as in a Salvia I saw. It grew in leafy clumps, from which rose tall stems covered with large pale lavender flowers. These shoots of bloom, which were sticky, reached nearly three feet above the coarse masses of foliage; a fine plant for the herbaceous border.

The flowers were not the only things which had changed. There were new birds, such as magpies and choughs; and new animals, such as marmots and hares. I was in a new world.[3]

That night, my first alone in new country, depressed me. I would have given a lot to have had my companions with me, and chatter idly after dinner. I turned to my botanical work as the best antidote to loneliness.

Next morning several of the coolies were snow blind and unable to move, which delayed the start. They lay in darkness groaning.

Just below our camp the torrent plunged into a gorge. As there was no way through it we had to follow a narrow path up the steep cliff. This path ascended for 1000 or 1500 feet, and then traversed the mountainside where the cliffs slanted back. On the screes and cliffs alpine flowers grew in abundance, especially the dwarf magenta *Rhododendron keleticum*, the vivid flat flowers of which were spread so densely over the matted stems that no leaf was visible. Milk-white Cassiope formed heather-like tufts.

There was a dwarf Cremanthodium on the bank of a stream.

On the Tibetan mountains there are many species of this genus, some of which occur in great numbers. Cremanthodium belongs to the huge Compositae family and is closely related to the universally spread genus Senecio, to which the English Groundsel belongs. It is in fact the Himalayan expression of Senecio. But whereas Senecio includes many dull or inconspicuous species, all the Cremanthodiums are beautiful and would be grown in our gardens, if we could grow them. This species was remarkable for its deeply-cut leaves and large, solitary, fragrant flowers, yellow, with a velvet brown collar of bracts.

From the cliff I had a good view of the range we had crossed, not to the Ata Kang La behind (i.e. south of us), but to the west. There was a barricade of snowpeaks, reinforced with many glaciers, flowing to a common lake. I caught the flash of other lakes in the hollows of the ice-planed rock. To the north I saw another spur down which half a dozen glaciers crawled.

Where the valley widened the path descended abruptly to the river again. There had been a lake here formerly, but it was now silted up. A herdsman, the keeper of the yak we had seen the previous day, had pitched his tent in a meadow.

Rhododendron sigillatum grew densely on the cliff we had just come down. Many of the bushes had rose-pink instead of white flowers. I refer to Rhododendrons at length, because they are an index to climate. Though at Chutong Camp I found twelve or fifteen distinct species, from the village of Lhagu northwards throughout the country I traversed, I found only four species, three of which I had not seen south of the range.

A very clean glacier descends almost to the plain at this point. As we crossed its outflow stream we had a good view of its steep snout. The North Ata Kang La glacier, which is the largest I found on the north face of the range, stops at a little above 14,000 feet, and no glacier that I saw descends lower than about 13,000. On the south side it will be remembered the ice foot was as low as 8000 feet.

The torrent now entered another gorge, but was not shut in by towering cliffs. The path left the torrent and ran over undulating hills, from the last of which we got our first glimpse of Lhagu, houses and fields scattered down a gentle declivity, sloping towards a lake. On a hill was a small square stone temple, surrounded by gnarled junipers; below that a dozen houses. The boggy pastures along the foot of the mountain were brilliant with yellow Primulas and crimson Lousewort (*Pedicularis*). The stone walls round the barley fields were piled with brushwood to keep out the cattle; but blue Cynoglossum threaded through the dead thorn, and Clematis trailed over all. Among the many flowers that grew round Lhagu, as round all fertile villages, were Asters, Astragalus *Gentiana barbata,* Trollius, and violet Geranium.

A large glacier from the north-west blocks the lower end of the Lhagu lake. The river running underneath its snout has undercut it so much that small icebergs are continually flaking off, leaving the ice cliff gorgeous blue with a bottle-glass fracture.

CHORTENS AT SHUGDEN GOMPA

At Lhagu we obtained baggage animals and riding ponies once more. The path to Shugden Gompa ascends a cliff opposite the glacier, and from the top of this cliff the white buildings of the monastery are visible for the first time, about twelve miles distant at the north end of Shugden Lake. The Lhagu Chu here, leaving the gorge, spreads across the mile wide valley, in many interlinking streams. At one time the Shugden Lake reached the foot of the Lhagu glacier, but it has now been partly silted up. The path presently descends from the cliffs to the valley floor.

A high pyramidal peak, called Dorge Tsengen, rising as it seems from the lakeside opposite Shugden Gompa for 7000 or 8000 feet, dominates the valley.

For two miles we rode over the flat, hard-floored valley, dotted with low bushes of Hippophäe. Then crossing the Zo La Chu, which flows in from the south-east, we halted for lunch at the village of Yatsa.

The day had been fine so far, but not warm: though it was not too cold for me to eat my lunch sitting on the turf. We had to change pack animals here and the people of Yatsa suggested I should stay the night, saying their ponies were out grazing. It was not yet 1 o'clock and I was anxious to reach Shugden Gompa which was only another eight miles, so I told Tsumbi we would go on with the few ponies available and the kit could follow next day.

Below Yatsa the present lake begins; and the path mounts the cliff once more and keeps along a high narrow shelf for some miles. Now with the last trees left behind on the edge of the snow range, and the plateau unrolling before me, I realized for the first time into what a sterile land I had come. Large areas of cliff were barren of plants, just bare rock. There were no trees and few shrubs. The distant ranges looked barren and hopeless. What plants there were, were stunted; not small merely, but uncouthly disproportioned, their leaves being squashed into rosettes or bunched in cushions, hard and smooth, with large short-stemmed flowers. What sacrifices plants seem to make for their flowers!

It would be easy to attribute this stunting to wind, the short season, drought or cold, and no doubt these factors play their part. But as I looked at the barren slopes, I wondered whether the dwarfing of plants was not due more to starvation, to a lack of nitrogen in the soil. This seems to me possible, even probable. It is a matter of common observation that alpine plants from Tibet often grow larger and are better proportioned in this country than in their own. Dwarf plants cease to be dwarfed; and the reason may be that they receive a full ration of nitrogen.

The plants that occurred in greatest numbers in dry ground were *Stellera Chamaejasme*, Delphinium, Astragalus and Pedicularis species; in moist ground *Primula tibetica*; by running streams *P. sikkimensis* and yellow-flowered Cremanthodium. These plants were obviously best adapted to local conditions, to poverty of soil as well as to a severe climate.

Near Shugden Gompa the country opened out and the steep cliffs gave way to rolling hills. We rode along the shore of the lake and then turned away and came to a few scattered houses; cattle grazed in the hollows. At last we returned again to the lakeside and having crossed another river, which had thrown out a large delta, constricting the lake, we reached a village encircled with cultivated fields. The monks had pitched a group of white canopies among some decrepit cloisters or shrines.

When they saw me they came over and surrounded my pony. They wore red robes and their heads were shorn. They looked at me gravely with faces like Buddhas, silent and without emotion.

We passed through a crowd down a lane between hedges of piled thorn as far as the foot of the steep ridge. An ascent of some hundreds of feet brought us at last to the monastery outbuildings. Passing the temple and several more shrines, we reached the fort, where the Dzongpön or district officer resided. We rode through the wooden gateway into the yard and I was shown upstairs into a dingy room on the first floor.

The time was five o'clock. The sky was overcast and a keen wind blew. I felt cold, tired and depressed. The valley's unexpected bleakness disappointed me: and my reception was anything but warm. My only greeting had been the incurious stare of the monks, who for the first time were seeing a white man. I hate being an exhibit.

There was not even a fire in the room to welcome me, though Tsumbi soon put that right. I missed the company of Ronald and Brooks-Carrington. My room was dark and draughty. It was approached through the kitchen. It stank. I told Tsumbi that I would not stay in that room. He must find me another. He went away to complain to the Dzongpön, and was gone some time. As last I undressed and went to bed.

CHAPTER X

THE LONE MONASTERY

Next day I awoke to my noisome cell, feeling all the better for a good night's sleep. I had hardly finished breakfast when Tsumbi came in and said that the Dzongpön had found another room for me, if I would care to change.

He took me along a dark passage, past a door, and crossing a small inner chamber, we came to a large room in the corner of the building. It was quite empty and a man was sweeping it. It had obviously not been used for some time. It was three times the size of my old room, and quite well lit by two windows looking over towards the monastery. Beneath a square hole in the roof stood a movable fire box of wood and beaten earth. The outer walls were of mud, about a foot in thickness; and the roof, which was rather low, was also of mud and withies. The floor was boarded over. I was pleased and said this room would do me well, which reassured the Dzongpön, who was afraid that he would lose me. The dzong was a grim looking place. A dark low passage round the corner from my room ended in an *oubliette*. On each side there were two latrines, dim rooms built in the thickness of the wall, opening below into a sort of midden, where pigs and mangy dogs scavenged. A ladder led up to the flat roof from which I had a perfect view of the lake and surrounding country.

I had my boxes brought in and arranged along the four walls. My table and chair were set up by the window, with the camp bed alongside. Field-glasses, compass, thermos flask and camera were hung from pegs. Lastly my flower presses and stacks of drying paper were piled in the centre of the room against a square wooden pillar, supporting the ceiling. The bare room now looked habitable, or at least inhabited.

The day was bright and sunny, but the teasing wind persisted. I found indeed that this wind was almost constant by day, and that when it dropped, rain fell.

THE LONE MONASTERY

I felt so light-hearted that I determined to climb the hill, which sloped up behind the monastery to a conspicuous ridge, and take a look at the country.

The monastery straddled a low isolated ridge which ran parallel to the lake for half a mile. To the north it descended to a flat plain where the village stood. The dzong adjoined the monastery at the southern end of the ridge, which ended in a cliff, at the foot of which a river, two hundred feet below, rushed from the north-east through a limestone gorge. A shallow dry trough separated the monastery ridge from the mountains behind. This trough was sheltered from the rasping wind, and its banks were covered with shrubs,[1] among them *Rosa sericea* and another species of rose with pinkish or cream coloured flowers, usually both on the same bush, which grew to flask-shaped fruits of a scarlet colour: Cotoneaster and a currant whose leaves turned wine red in October; a Barberry of no repute and a wizened form of the shining leaved *Lonicera Webbiana*, forming stout bushes. There was even a grove of scraggy juniper: and at this season, bushes were trailed with long ropes of Clematis knotted with rich yellow flowers, which linger, darkening to mahogany, long after the fruit begins to ripen.

This trough-like valley which led down to the lake was my favourite place to stroll in the evening and early morning. There were always hares to be seen here and flocks of chattering babblers under bushes. On the bank I found the large pearl-white flowers of *Anemone rupicola*.

From the dzong a broad road crossed the trough, skirted the main ridge and disappeared behind a shoulder in the direction from which the torrent came. I surmised that this was the road to the Salween, but it was also the main road to Lho Dzong and Lhasa.

This road I followed for a short distance, then turned off and climbed the dry slope on my left, making for the crest ridge. Herbage here was scanty, though a thousand clumps of sweet scented *Stellera Chamaejasme* shook in the wind. Its large woody rootstocks are proof against drought.

Though the crest of the ridge had looked barren from Shugden Gompa, I found many flowers here: blue poppies (*Meconopsis horridula*) and the flamboyant purple *Incarvillea brevipes* which looks as if it had escaped from the tropics. The turf, a brilliant green in some of the dips and hollows, was spangled with orange hawkweed. Anemones and Asters covered the sunny slopes.

Climbing to the highest spot I could see above me I had a view of the surrounding country. The ridge on which I stood was divided from the main mass of eastern mountains by yet another valley with a plain at its head, which ended on the cliff overlooking the deep river gorge. This plain—an obvious silted-up lake basin, called Bütang, was grazing land on which strayed a few ponies and cattle close to a small village.

Eastwards, the ridge fell steeply away to the narrowing valley, down which a stream flowed, overgrown with bushes on its bank: while to the west the slope was gradual towards the lake, though at the north end where the ridge rose to a peak, marked by a cairn, there was an abrupt cliff. Southwards, looking towards the Ata Kang La, I saw a great range of snow peaks and glaciers. Two gaps showed the roads into Zayul, one to the east in the direction of Sangachu Dzong, the other towards the Ata Kang La. North and east the view was soon shut off by high crags of slate and limestone: but though there was much snow, I detected neither permanent snow peaks nor glaciers.

In the west the view was obstructed by the high range which rose sheer above the lake. 'A. K.' and Bailey had both reported the lake to be some four miles in length. From where I stood I saw that it was much more than four miles long. It was narrow like a fjord and though I could see the northern end, it took another twist to the west at the foot of the mountains and soon went from sight. These mountains, soaring from the lakeside, reached their zenith in Dorge Tsengen, about 20,000 feet in height. Shaped like a pyramid, it is so sheer that no snow lies on its south face. A glacier descends the north side towards the lake, but stops short in a valley before it reaches the water.

Dorge Tsengen is a landmark from the south. Its tip is visible from the Zo La (as the pass to Sangachu Dzong is called) and even from the foot of the Ata Kang La glacier. Northwards, however, it is soon hidden from the road.

Across the lake, a row of corry glaciers hung poised almost in mid-air, their snouts truncated. While to the north-west two canine teeth of limestone stuck up into the sky. High as I stood, the country round was higher, the mountains to west and south-west being the highest of all.

The valley from the upper lake at Lhagu to the lower at Shugden Gompa had been scoured to a great depth; but for that, the country resembled an undulating plateau edged with high mountain ranges.

THE LONE MONASTERY

SHUGDEN GOMPA

Ice and water had worked on the plateau; enormous glaciers first ploughing broad furrows across the country—it was these had brought from distant ranges the immense granite blocks I noticed on Bütang and set them resting on the upturned edges of limestone strata. Of these erratics, one I saw was fifteen feet in height and more like forty in girth.

The glaciers, then retreating, left long lakes, which shrank in their turn and silted up. Even while the glaciers retreated, streams from their melting snouts scoured through the moraine deposits deeper and deeper into the plateau. The long isolated ridge on which I stood and its satellite, the monastery ridge which I could see far below, must once have been surrounded by glaciers from which perhaps they stood out like rocky islands. Later they were islands in a great lake which drowned the valleys.

I imagined these deep valleys filled in—there were only four of consequence round Shugden Gompa, all of them source streams of the Nagong river; from Ata Kang La, and Zo La to the south; Traki La, and Pö Yu La to the north. The outflow of the Nagong from the lake to the north-west made a fifth. If these valleys didn't exist, the country would be again a plateau land.

But such changes were superficial, caused by ice and water. The fundamental scaffolding of the country, the foundations of the mountains, were of sedimentary rock, and I found that Dorge Tsengen itself was formed by a broken limestone syncline, the strata curving from north to south. About Shugden Gompa was hard grey limestone, alternating with slate and similar sedimentary rocks; whereas between Rima and Ata, the rocks are invariably mica schist, gneiss or granite. Away to the south-east towards Zo La was a mountain of bright red sandstone, evidently manganese-stained, probably of Devonian age. Limestone and slates or shales ran in bands right across the valley from Dorge Tsengen towards the Salween, the rocks dipping at high angles (75°-90°) so that only their edges were visible. The main limestone bands, lying between two bands of slaty shale, had a thickness of not less than 10,000 feet. Thus the strike of the rocks was invariably north-west to south-east or west-north-west to east-south-east. Almost certainly they had not been beneath the sea since the Primary epoch and were of Devonian or Permian age perhaps.[2]

Having observed my surroundings and fixed on Bütang as a suitable place on which to measure a base for a map, I collected some plants and returned to the dzong.

That evening I completed settling in, getting out my books, instruments and such clothes as I should require for the time being. I realized that while it was pleasantly warm by day, I need not expect summer temperatures of much over 60° F. At night, the temperature sank well into the forties. If it was no hotter than that at Shugden Gompa (13,500 feet) in July and August, camping at 15,000 feet in September and October was going to be cold.

The south wind blew all day, sweeping off the glaciers down the long lake valley. It cut like a whip across my face as it scudded over the ridge I loved to explore. I feel it now, fluttering the prayer flags of the monastery, gusting down the chimney, filling the room with smoke. It blew ceaselessly without pity, till it got on my nerves and ruined my temper. I grew to hate and fear it because there was neither rest nor shelter from it. It was a constant, nagging enemy: but in autumn, when I was collecting seeds of Gentians, it seemed right, tonic and friendly.

Little rain fell. I could see that from the clear atmosphere and the harsh jagged outline of the mountains. Nor could there be much snow, either, though the winters are iron hard. Day after day I watched the waves of southern cloud break on Ata Kang La, and saw the cloud spray ripped to shreds over our side of the barrier by the wind.

Most days in July were fine and there was lots of sunshine—a relief after six weeks' living in the clouds. Perhaps 1933 was a specially dry year: for I find it difficult to believe that even an alpine flora, which needs little rain so long as it is supplied with underground water from melting snow above, could thrive on such drought rations. A species of Morina, thistle-like in appearance, but actually a teazle, was in bud when I arrived at Shugden. But it never opened a flower: not a seed ripened. The whole mass withered. That argued drought conditions. But as on the same slope tens of thousands of other plants flowered and seeded—I recall especially the violet-flowered *Dracocephalum tanguticum*, whose dried fruit-heads smell of verbena—the verdict is not proven.

For fifteen days in August it rained rather heavily, but September was fine on the whole; also October, when snow began to fall. One might hazard that on this part of the plateau the annual rainfall does not exceed twelve inches, of which most falls between June and September.

At Shugden itself, rain nearly always came from the north-west, from Pomé and the Tsangpo: but of course I often stood in sunshine, watching storms circle over the distant ranges, though no rain fell in the valley.

The second day after my arrival I had visitors. First came the Dzongpön. Though it was an official visit, he was not very well dressed. He had put on a dark wool *chupa*, but had only put an arm through one sleeve. I noticed that underneath he wore a dirty shirt, buttoning on the left shoulder. His black leather boots into which he stuffed his baggy cloth trousers were the smartest part of his dress. He was a typical Kampa between fifty and sixty years of age, with long untidy hair and straggling beard, and a round cheerful dirty face.

He seemed torn between duty and politeness; and was nervous and hesitating at first. He asked if I had a pass and would I show it to him? He apparently knew nothing whatever about me.

I was surprised that the Lhasa government, who had given me permission to go to Shugden Gompa, had not warned their local officials. But it was no great matter. I seated the Dzongpön on a box, gave him a cup of tea, and handed him my pass from the Kashag.

He read it through aloud, very slowly, as though letting its full weight sink in. Then he handed it back to me with a bow and remarked that it was quite in order as far as Shugden Gompa. Perhaps he had already heard from Tsumbi that I wanted to go to the Salween.

Anyhow, I lost no time in telling him, so that he might think it over. It was plain he had no idea what I was doing, or why I had come to this country. So I told Tsumbi to enlighten him.

My next visitors, however, were the abbot and several of his monks. They knew what I was doing. One brought me a bunch of flowers, which he had plucked on the ridge. I was familiar with most of them—a dwarf Iris, *Paraquilegia microphylla*, a purple bog Orchid, a globe flower (Trollius). But there was one I had never seen before, a Primula[3] with bright rose red flowers, deliciously fragrant.

I asked him where he had found it and he pointed out of the window to the ridge behind the monastery, where I had been climbing the previous day. I decided to go there again next day and collect it. I put the flowers in water, but as they were already rather withered, I did not press any as specimens. In fact, the only one I wanted was the rose Primula and I could easily find that for myself. So I thought.

The monks used to come into my room and watch me working. None of them had seen a white man before and they eyed me with curiosity. But they were always smiling and friendly. My microscope rather baffled them; they found it too difficult to focus, and they did not close one eye. The thermometer and compass interested them, but field glasses delighted them more than anything. After holding them away from their eyes, upside down and wrong end on, they finally got the idea and managed to manipulate the focusing. Seeing the landscape leap up close, so that they felt they could touch it, and looking at one another's faces enlarged, gave them astonished pleasure.

Then they would go and watch Tsumbi in the kitchen, seated cross-legged on the floor, changing the plant paper. They could not understand how I found so many flowers. They didn't see them, they said. I answered that they didn't look for them.

The Dzongpön was more suspicious. He wished to know why I should want to go to the Salween. There weren't any flowers there, he said.

CHAPTER XI

HIGH LIFE ON THE ROOF OF THE WORLD

The white-haired abbot asked me to lunch at the monastery. He wanted me to go, when they had their meal, at about 11. But Tsumbi told him that I never ate at that hour and arranged for me to go at 2 o'clock instead. This monastery, built of mud, stone and timber, was a much more substantial and magnificent building than the wooden monasteries we had seen in the forest country. The monks lived in rooms instead of in hutches.

When I entered the stone paved courtyard, I was met by several monks who bowed and led me silently down a dark passage. A grey greasy curtain was lifted and I went into a small, dark room, where a low stool or bench was set before a divan.

The abbot welcomed me and told a servant to bring food. He himself had taken his meal earlier and sat in an adjoining room talking to Tsumbi while I ate. I sat cross-legged on the cushion. The food was served in small china bowls and servants having placed chopsticks by my plate, at once withdrew, all except the tea man.

I began with a cup of reddish buttered tea, the national social lubricator. The man stood at my elbow, swilling the horrid wash round and round in a black teapot. I took a sip. As soon as I set down the cup, he filled it to the brim. Politeness whispered 'Drink'. Prudence shouted 'Don't'. As it grew cold, a film of butter coagulated on the surface.

One bowl contained rice, another gobbets of mutton, a third a spinach-like vegetable, a fourth the same vegetable pickled and very sour. I finicked with these drab delicacies with a pair of chopsticks; as each bowl, not much larger than a teacup, was finished it was replenished. At last I struck. The cold mucilaginous spinach defeated me. I refused further food and presently the abbot and Tsumbi came in to talk. Amongst other things he told me that Shugden Gompa was under the patronage of Ganden Monastery near Lhasa. He remembered

Bailey coming twenty-two years before; Bailey had slept a night in the monastery, but the abbot himself had not seen him.

After an exchange of courtesies, during which the abbot gave me permission to take any photographs I liked in the monastery, I returned to the dzong.

The next day, not to be outdone, the Dzongpön asked me to lunch. I went again at 2 o'clock, and, as before, after being greeted by my host, I was left to eat in solitude. This meal was superior to the one I had received the day before. In addition to all that the monks had given me, I had *chang* to drink, and *tsamba* rissoles filled with chopped meat and red pepper (capsicum), Chinese brick sugar, and dried yak flesh as fibrous as asbestos and about as edible. I was waited on by the Dzongpön's old servant, a reverend baldheaded, white-bearded man with a high deeply-furrowed forehead. So wrinkled was he that I began to think he had been made up in the green-room; could any man have so high and white a forehead, or was it a property wig he wore? His dingy coat clung to him in greasy folds; and he looked as if he had not undressed, even to sleep, since he reached manhood.

When I had finished, the Dzongpön came in and was affable. But we did not mention flowers or the Salween. I soon went up to my room, and having changed out of my best clothes, went for a walk. The depression between the monastery ridge and the next ridge was, as I said above, filled with bushes and wizened juniper trees. The main ridge, called Ningri Tangor, or rather the exposed side of it, was covered with flowers. Few species were represented, but those in great quantity. A pink Rattle (Pedicularis) and a yellow were abundant, as also a small single flowered Aster, like an Ox-eye Daisy. The far side of Ningri Tangor was thickly covered with shrubs and in places with good sized Fir trees. The plain of Bütang was simply pasture at the upper end; but where the ground moistened, it was starred with a pink pattern of the minute *Primula tibetica*. There was a pond here, in which was a large number of Gasteropods like the English water snail (Limnaea). Further north, where the plain narrowed and the seeping water began to form streams, a dense growth of thorny shrubs (mostly Hippophäe and Lonicera) and herbaceous plants replaced the pastures of Bütang. The main road to the north followed this valley, which ran parallel to the lake.

Among the Hippophäe thickets along the stream, I found a tall violet-flowered Geranium, a larkspur of brilliant blue and the dingiest Monkshood I have ever seen, its flowers a muddy amethyst. Thousands of Asters grew here, an ivory Grass-of-Parnassus (*Parnassia nubicola*) and golden Trollius with trilobed leaves.

It was a place where one might expect to find a Bog Iris. But the only Iris I found at Shugden Gompa was a dwarf alpine, bearded and violet in flower, with a capsule which split down the carpels, but never quite opened, the carpels remaining united at the apex.

However, I found one plant of unusual interest in this marsh—and nowhere else. This was *Ajuga ovalifolia*, one of the few plants which was both rare and local at Shugden Gompa, for most plants grew in millions. But there were scarcely a hundred specimens of *Ajuga ovalifolia*, and they all grew together on a few square yards of marshy turf beneath bushes of Hippophäe. I chanced on the Ajuga as I was forcing my way out of the marsh through the dense thorn scrub, which pricked me severely. It was in full bloom or I might not have noticed it; for it stood barely three inches high.

The basal leaves, which lie flat on the ground, have a spread of six inches. The succeeding pairs, decussately arranged, are smaller and the internodes telescoped. The bright violet flowers appear to form a compact head at the apex of the leafy pyramid; for those of the lower whorls have such long tubes that they draw level with the uppermost. Thus the inflorescence forms a brilliant cap, crowning a plinth of green leaves. Neither flowers nor leaves are scented.

I found this plant on July 20th. Six days later, having looked at the dried specimen of the plant and seen that it was good, I returned to the marsh to make sure where it grew and to mark the place, in case it was not a common plant. I had some difficulty in finding the spot. But when I did, I built a cairn on the path a hundred yards away in line with the Ajuga patch and a cliff across the valley; and then I built a second cairn by the patch itself. Thus, walking along the path, I could turn off into the thick bush and know that by keeping straight on a hundred yards I should find the spot. I rarely find it necessary to mark a plant so accurately. But I found no other place where this plant grew and dared not risk guessing wrong.

Its seeds were ripe long before snow came. My enemy was slugs, not weather. Had there been thousands of this plant, it would not have been serious. But as it was, three capsules out of four went to feed the slugs. Between August 19th and September 5th I went five times to the patch of Ajuga—for one good plant is worth a dozen inferior ones—and each time I secured a few seeds. The trouble was to collect the seeds before the slugs ate them, yet restrain myself from picking them before they were quite ripe. For unripe seed cannot survive the journey to England and then germinate. By the end, I had a few hundred seeds, about the size of mustard. They reached England five months later and germinated well in May 1934.

Flowers were not the only attraction in this marsh. Going from flower to flower were insects and butterflies, especially on sunny days. In alpine valleys, they ascend as high as flowers grow, to 17,000 or 18,000 feet. But naturally they were most abundant between 13,000 and 15,000 feet and Bütang was a favourite place.

The butterflies were mostly small and rather dull coloured, more like our British butterflies or those of the temperate Eurasian continent, but none the less interesting for that. In all there were twenty or thirty species common on the marsh, among them a Tortoise-shell, two Clouded Yellows, and several Small Blues.

Grasshoppers abounded, but I noticed only two varieties. Grasshoppers are more significant than their size would lead one to think. To the naturalist no plant or animal is so trivial, that if he knew its exact distribution in space and time, it would not throw light on geological history. The power of locomotion that lower forms of life possess is very limited; and the discovery of similar plants or animals in different localities may go far to prove an unsuspected land connection of great antiquity between two areas, while it shows definitely the direction in which a certain flora or fauna migrated. Thus my discovery of two new species of grasshopper, belonging to the genus Podisma, on the Assam frontier some years ago, enabled B. P. Uvarov to state categorically that there was a Palaeartic element in the grasshopper fauna of the eastern Himalaya, which connected that region with China (the Angora continent) in preglacial times; and further proved that the history of the eastern Himalaya must be very different from that of the western end; a conclusion which botanists will endorse.

I found green striped toads in the streams which flowed through the marsh; but owing to the absence of forest there was little mammalian life round Shugden Gompa. Except for hares, which swarmed in the trough below the monastery, an occasional pigmy hare (Ochotona) on rocks at higher altitudes, and marmots, I saw no animals here. But I saw plenty of birds; orange-beaked choughs, rose finches and a large black and grey rock pigeon being the commonest. There were several species of babbler; one yellowish-brown, another long-tailed and reddish-brown with a song like an English blackbird's. The babblers migrated after the first week in October. A pair of white-tailed eagles often soared over the lake, and I saw an occasional jack snipe along the shore. Though once or twice I flushed a solitary snipe, the only common game bird was a small partridge of which I put up several coveys in the Rhododendron scrub. I saw several magpies and once a hoopoe.

Below the monastery, on the delta thrown out by the Tzengu Chu stood a small village, close to the lake shore. The delta divided the lake almost in two, the two sheets of water being connected by a channel about a mile in length. Most of the delta was marshy in summer and given over to grazing cattle, ponies and yak, but near the village it was dry and cultivated with fields of barley.

The lake has a total length of some ten miles, if we include the two connecting channels where the silt brought down by streams has encroached. It is nowhere more than a mile wide. The Lhagu glacier, swollen by other glaciers, must once have reached to beyond Shugden Gompa, and is largely responsible for the excavation of the valley.

I saw a few tiny fish inshore, and there may be larger ones in deeper water, but no one troubles to catch them. The shore line is remarkably regular, but there is no flat shore for villages and cultivation except where large streams enter. The lake level varies considerably. Much of the water is shallow, and the lake shrinks, as the glacier streams dry up in autumn, till by October most of the grazing land below Shugden Gompa, which had been under water in August, is dry land again. The Tzengu Chu would become turbid in a night and run with mud for 24 hours, then turn limpid again quite suddenly. In the absence of steady rain, sudden storms smote the high peaks, which likewise turned white during the night, but the snow film disappeared in the course of a morning.

The only crop grown in Shugden is barley, and when this is reaped at the end of September, the fields remain bare for some months. I was able to buy *tsamba,* made from this barley. It is the only cereal product of the district. Wheat-flour and cornflour are imported from Zayul. I succeeded in getting a little from the monastery, which is, as always in scattered communities, the local stores. The monastery imported rice also in small quantities.

The only fresh vegetables I could get were peas (in August) and a few turnips. The Dzongpön's wife supplied me daily with milk and butter, but there were no fowls and so no eggs at Shugden Gompa. As for meat, I bought a whole sheep on my arrival—a tough ram it was too; and from time to time I purchased a little yak flesh, or a goat or a pig. Yak are not killed; they die.

There was no water on the monastery ridge. It had to be fetched from the river below in wooden tubs which were laden on to yak. Three of these beasts worked all day long, plodding back and forth to the dzong and monastery, each carrying two great tubs of water, in which a handful of twigs were placed to prevent it slopping.

The people of the plateau are Kampas, which means simply 'inhabitants of Kam', the name given vaguely to Eastern Tibet. They are taller and finer featured than the people of Lhasa, though they are evidently of mixed origin. Few of these Kampas have ever seen a white man before, but having long had contact with the Chinese, they are broader minded and more progressive than the people of Central Tibet. They are naturally hospitable and friendly to strangers, though they are liable to outbursts of fanaticism.

Though the severe climate of Shugden is not favourable to good looks and still less to good complexions, many of the red-cheeked girls I saw in warmer valleys were quite pretty. Women do their hair in a peculiar way. After buttering it, they plait it into a number of thin rats' tails. Then these are gathered together at the end and plaited in with blue wool till the thick queue trails almost to the ground. They wear a spindle-shaped silver brooch on top of their head, with a coral in the middle and a turquoise at either end.

Shugden Gompa is the centre of a district called Nagong, stretching from the Ata Kang La and Zo La in the south to the Poyü La in the north, a distance of about forty miles; and from the Traki La in the east to the west end of Shugden Lake (about twenty miles). Thus it actually comprises only the four headwater valleys which discharge into the lake, from the west end of which the Nagong river issues. These valleys, which vary between 12,000 and 14,000 feet above sealevel, are a small proportion of the whole district, but the only part which is habitable.

There are some twenty villages near Shugden, of which more than half are scattered round the shores of the lake. They vary in size from two or three to possibly a score of houses. Taking their average as eight, with five persons to a house, we arrive at a total population of 800, to which might be added 100 monks. For the whole area of Nagong, some 800 square miles, the population is certainly not more than 1200, and of these many are children who will die before they reach maturity of disease brought on largely by malnutrition.

This scattered territory is administered by the Dzongpön, who, however, rarely leaves the dzong. Shugden is a second-class dzong, and the Dzongpön, unlike the Governor of Zayul, is not directly under the Chamdo administration, but under Pashu,[1] a first-class dzong some five days' journey to the north. A first-class dzong is usually in authority over several of the second class: Pashu, for instance, has six or eight satellites.

The duties of a Dzongpön seem light. He maintains order, but there is no disorder. He settles disputes, but there were no disputes while I was there. He is a magistrate with no cases to try. I never knew him to receive or dispatch a letter to his superior, though he told me he had received verbal instructions (brought by a caravan leader!) from Pashu to treat me well! His real task is to collect taxes and check tea, salt and other caravans which pass through Shugden in the autumn and change transport there. After three years' service in a second-class job a Dzongpön goes to Pashu, and is then appointed—and possibly promoted to a new district.

The abbot seemed more important in the eyes of the people than the Dzongpön. His monks were as active in their pursuit of business as the Dzongpön was slack—though that was more the system's fault than his own. Day after day I saw the monks in long cloaks and white cloth sugarloaf hats ride off on quick trotting ponies to distant villages, or nomads' tents in the mountains. And in the evening I heard the jingle of bells as the party returned from its ministrations.

The nomads' tents, made of black horsehair, are pitched at between 15,000 and 16,000 feet. The villages, as well as the monastery, own large herds of zo (cross-bred between a yak and cow), cattle, ponies, sheep, and goats, which they send up to the highlands in summer under the care of these wandering Ishmaelites. Grazing rights are strictly

GLACIERS ON THE GREAT SNOW RANGE

apportioned, and every alpine valley belongs to one or other village. These herds do not return to the warm lower valleys till mid-October.

I decided it would be better policy to settle down at Shugden Gompa for a time and make friends with the Dzongpön before I again suggested a trip to the Salween. Besides, I was anxious to explore the immediate neighbourhood of my base, since it would be easier to collect seeds here than further afield, and I wanted to make a plane table sketch of the district, for which purpose I measured a base on Bütang and fixed a few of the more prominent places.

Every day I went out plant hunting, usually on Ningri Tangor. The day after the monk brought me the rose Primula, I hired ponies from the Dzongpön and with Tashi and a coolie rode to the top to get it. But though I searched the grassy slopes, which were gay with violet Anemones and other flowers, I found no sign of the Primula. A few days later I tried again, searching the sheltered wooded sides, but with no better result. The summit of Ningri Tangor was broken off towards the lake and sloped precipitously. The fluted cliffs had flung down screes of shale. It was no use looking on those bare crumbling slopes for a Primula. But all the same I went a little way down and below the scarps I found clumps of *Paraquilegia microphylla* and a remarkable Potentilla with long woody stems covered with the brown leaf bases of previous years. Some of the plants I saw must have been thirty or forty years old; though like dwarf trees, they grew an almost infinitesimal amount each year.

But there was no rose Primula and I began to wonder whether the monk had made a mistake.

CHAPTER XII

EXPLORING

ALL that I or anyone knew of the River Salween was that it was somewhere to the north of me, but I didn't know in which direction, whether south or south-east, it flowed. The accepted theory is that the Himalayan range was turned south-west at the Tsangpo bend, through coming in contact with the far older rocks of Western China, which run north and south, and that on these the Himalayan earth movements made no impression. From observations made on my earlier journeys, I had begun to suspect that this theory required modification; that in fact the Himalayan convulsion *had* made an impression on the pre-existing rock, elevating a snow range west and east, at right angles to the apparent north-south alignments. I thought that the orography of the gorge country had been studied too much from river courses, and too little from the alignment of mountain peaks, and that on examination these rivers would be found to have cut courses right through the real mountain ranges, instead of running parallel to them. Owing to the number of rivers piercing this range close together, the appearance has been superimposed that there are many ranges running north-south, instead of one long range running west and east; that is, the true breadth of one range has falsely been regarded as the lengths of several ranges, each river divide being regarded as a separate range. I was led to this surmise partly by the fact that all the snow peaks that I have seen during the last eighteen years, when plotted on a map, lie in a belt running north-west to south-east. The breadth of this belt nowhere exceeds seventy-five miles. Thus I did not regard the river divides as separate north-and-south ranges, but as isolated strips of one west-and-east range. Where the Tsangpo bends south, it passes through a deep gorge, as it would if it were cutting its way through a range. I hoped now to discover whether the Salween did the same.

Having got permission from the Dzongpön to proceed beyond Shugden, I made all preparations, and on July 30th started finally on what I was told would be a four days' march to the river.

The first day and a half were spent in a gradual ascent up to the Poyü La, a pass over the mountains in a northerly direction from Shugden Gompa. Just below the pass there is a lake called the Poyü Tso, which has no visible source or outlet. I was told that it was held to be sacred and that its waters ebbed when the health of the Dalai Lama was poor. There was plenty of water in it in July, and still plenty in October, when I saw it for the last time. I don't know what happened two months later when the Dalai Lama died. There were several nomads' tents here and large herds of cattle.

The ascent from Shugden was not more than 2000 feet in all; yet the range we crossed was the watershed between the Nagong and Salween rivers. The range of which the Ata Kang La was the pass, ran parallel to this range. I did not, and do not yet know, whether these two ranges converge or not; more likely they are not two ranges but two crest lines of one range. It is important to remember that while the descent to Shugden Gompa is 3000 feet from the Ata Kang La and 2000 from the Poyü La, it is 10,000 feet to the Rong Tö river and 6000 to the Salween, which suggests that Shugden is in the centre of a mountainous plateau between two crest lines. These two crest lines, the southern crossed at the Ata Kang La, and the northern at the Poyü La, are thirty miles apart.

The evening of the day we crossed the Poyü La we came to a large village called Drongsa (officially Shoshi Dzong), at the junction of two streams. It is set in a fertile plain, an oasis among the rocks. The monastery was large and solid and very white; it looked more like a fortress. Beside it the dzong stood, looking more like a monastery. From a distance, the village seemed prosperous, rich even for these parts. But when we came close, it proved almost desolate. The houses could have held 200 people at least, but while we were there we saw no more than a score. The dzong contained only an old caretaker, and when I asked if the Dzongpön was away, he told me there was no Dzongpön. Apart from two figures we saw looking over the battlements, the monastery also seemed to be unoccupied. I wanted to see over it, but I was warned by the caretaker that fierce mastiffs wandered loose about the courtyard; and when I walked past I heard them making a noise more like lions roaring than dogs barking. When I went back there in October the monastery was just as deserted, so that it cannot have been that the monks were riding out.

I spent the night in the dzong. After dinner I strolled on the flat roof, watching dark clouds sail across the moon, which was almost full. For all its richness, the small plain soon ran up to screes and jagged peaks, whose outlines were sharp in silhouette against the dapple sky. Moonlight on the monastery made it look medieval; and watching it, I realized that nothing European is closer to modern Tibet than the Middle Ages. Tibetan life and culture, dominated by religion and a rich ruling class, is purely medieval. The peasants can neither read nor write. They love singing and dancing, and hold festivals and religious ceremonies. They live amidst beautiful surroundings, and are certainly not unhappy.

In the morning I inquired how far it was to the Salween. No one knew. Everyone said something different, but all agreed it could not take less than four days. The Salween seemed to be receding.

We had to change transport here, it was the custom and the Shugden yak owners demanded to be paid off. I did not think I should get fresh men in this deserted village. But Tsumbi summoned the headman and gave orders for fresh transport, and as our requirements were modest, we were not held up as I had feared.

As we continued northwards the valley became more and more thickly populated, and there was one cluster of hovels after another. The main stream, as if by magic in this dry land, swelled to a torrent, which was spanned at intervals by wooden bridges.

We passed few people on the road except mounted monks, and they rode by without speaking. But I saw men working here and there in the fields. A flight of yellow Clematis trailed over a hedge, the thimble flowers tottering on slender stalks.

The people were very raggedly dressed, appearing much poorer than those at Shugden. The women chiefly wore sleeveless jackets of goatskins, the hair of which was turned in towards the body, less clothes than breeding grounds for lice.

After travelling seven hours down the valley, making at least fourteen miles, we came to Trashitze Dzong, built on a high conical mound in the centre of the valley overlooking the river and several villages. There were groves of Buckthorn in the sandy bed of the river, which broadened out at this point, the current slackening.

Trashitze Dzong is half-fortress and half-monastery, and the lama, a friendly white-haired man of sixty-six, also performed the functions of Dzongpön. In him, as is common in Tibet, Church and State were united.

On the whole of my trip to the Salween, I was quite unexpected at the villages where I arrived. Neither the Kashag at Lhasa nor the Dzongpön at Shugden Gompa had warned them I was coming, yet by all of them I was greeted with hospitality and helped on my journey. Perhaps the news had got round that though I was a friend of the government, I paid for service and supplies, and that pleased the people. Whatever it was, I was treated very well.

The lama had the best of the many empty rooms swept out for me and sent me a present of eggs, milk and green peas. I gave him a pair of dark snow-glasses for his weak eyes, which greatly pleased him, and a dozen empirin tablets, of which I had a large supply.

From Trashitze Dzong a road, leading to Lho Dzong and Pashu, climbs the mountains and continues north till it eventually joins the great road from Chamdo to Lhasa. It was up this road that 'A. K.' had gone. Our routes had coincided from Rima to this point and I found his route report and traverse admirable as far as Lhagu; after that it inexplicably became so confused, inaccurate and misleading, that I found it useless; from Shugden Gompa to Trashitze Dzong I could hardly reconcile my route with his.

I now branched off the main road, travelling through country which had never before been described; country for which the next traveller has only my uncorroborated testimony and route map as a guide. We still followed the river called Tsa Chu.

The lama told me that the next day we should have to change transport seven times, as it was customary to change at each village. Our destination was Gongsar Gompa down the river, which turned east at this point. I finally persuaded him to let me off with three changes and he called in a young monk who wrote a letter to the village headman to that effect. I watched him writing, sitting cross-legged on the kitchen floor. A messenger was then called and told to take the letter at dawn to the next village, so that the first change of transport would be ready on our arrival.

The next day was hotter and the lower we descended the hotter it became. The valley narrowed and deepened, becoming more of a gorge with every mile. The granite of the upper valley had changed to sedimentary rock, red sandstone and black shale, with conglomerates

RAMBU GOMPA

at the base of the mountain. Every stream entering the valley from the north was so reddened with sandstone that it was the colour of cocoa.

In the afternoon black clouds piled up in the east; and as the landscape darkened, it was transfixed and lit with vivid shafts of lightning. Soon torrents of rain began to fall with extraordinary ferocity. Apart from the darkness cast by the storm clouds, the air was so thick with rain that it was hard to see a few yards ahead. The streams rose with amazing speed, flooding the main valley.

I soon realized that we could not reach Gongsar Gompa that night. At five o'clock we came to within sight of the third-change village, but found ourselves cut off by a torrent in spate. We were separated by a deep ravine down which ordinarily a feeble trickle flowed. But this had risen and was now flowing with tremendous violence. It was flimsily bridged by three logs not more than twelve feet in length, and ordinarily one would have crossed in four strides and thought nothing of it. But now it was different. The banks were slippery and the torrent was so high that it threatened to wash the logs away. I heard the muffled noise of great rocks smashing and grinding one another, though I could see nothing through the boiling cocoa-coloured water. It was still pouring with rain and we were wet through. We began to shout.

We shouted and shouted until after a time the villagers came to our help. They threw a rope across to us. Men, held round the waist by this rope, were able to stand on the flimsy bridge and pass the loads across. By the time that we had crossed, it was dark and we halted at the village for the night.

For two days after the storm the torrents ran mud. There was no clear water anywhere. We drank water the colour of cocoa, and our tea looked like cabhouse coffee. I don't know what coffee would have looked like.

The red sandstone was like the Devonian sandstone of Yenching on the Mekong. There were black shales, mud stones, slates and conglomerates in this part also, and we came to something resembling coral limestone.

When we rode across a mud fan spread over the valley, I had an experience of how these mudstones are formed. From a distance the grey fan looked hard as slate, but our horses sank to their fetlocks in it. The side ravines descending from the plateau rush with mud for twenty-four hours and then are dry for as many days. In spate they hurl rocks and debris of every kind into the valley fanwise, arranging it according to its size. And afterwards the silt hardens.

What doubts I had of the arrangement of the rocks was dispelled next day, when I saw red sandstone and white limestone strata highly tilted, forming parts of broken arches.

Gongsar Gompa was a picturesque building, similar to Shugden and surrounded by many scattered houses. All the corn had been reaped, and the roofs of the houses and even the monastery were stacked with sheaves of golden barley. The villagers were flailing the ears and singing as they worked. Others were ploughing. Platoons of pigeons strutted through the stubble. It was high harvest here.

We halted two hours waiting to change transport, and I rested in the headman's house. A crowd gathered in the small room to get a good look at me. They had never seen, possibly never heard of anything like me before. But they were good tempered about it. The heat grew so great and the flies were so annoying that I went to the window. From it I saw several pretty women, who appeared to enjoy being looked at, and who smiled at me.

These Kampas have window boxes of Tropaeolums and Petunias, which might have been and probably were, raised from Sutton's seeds (these plants are anyhow of S. American origin, not Asiatic), and in wooden tubs made from hollowed tree trunks were Sunflowers and Hollyhocks. It amazed me, who had come to Tibet to get flowers, to find the Tibetans cultivating flowers common in English gardens.

The people brought me peaches, which though edible raw, are better cooked, and a man brought in three skins of snow leopard, which he wanted me to buy. He wanted R.40 each for them and though he would have come down in price, they were not good enough for museum specimens. The claws had been cut off, and much of the head; the tail had not been taken out, and of course there was no skull. They sell them usually in Tatsienlu or Lhasa, whence they are exported to Shanghai and Darjeeling. I also noted the skin of a snow cock, rather moth-eaten, hanging in the room.

As we continued our march, the valley grew even more populous, with villages on either side of the river, which was spanned every few miles by elaborate cantilever bridges. On a white limestone outcrop I read the well known 'O MANI PADME HUM' (Oh! Sacred jewel in the lotus) carved over and over again. At another spot the rocks were bent and twisted into fantastic zigzag curves.

We passed more traffic than usual; swarthy men from Tsarong with donkeys carrying bags of salt, the slowest form of transport as I know from experience. When ponies are used, foals, accompanying their dams, added to the confusion without lightening the loads.

Though we had been marching five days, doing twelve to fifteen miles a day, there was no immediate prospect of reaching the Salween. We camped at Aule for the night and I got an observation for latitude for the first time since leaving Shugden Gompa.

When we reached Watak on August 4th the whole population came into the narrow fly-infested street and clamoured to do the next stage in my company. Men, women and girls seized on my boxes, each taking one article, even if it was no more than a camera. And seventeen in all distributed the loads between them. Even an empty biscuit tin, which I used for plants, had a coolie assigned to it. As usual, the heaviest loads were given to old hags, presumably on the principle that being of no further use to the community, they might as well die in harness. The girls were dressed in patchwork skirts made of such thin rags I expected they would flake to bits as they walked. Yet they all wore fine chromatic cloth boots.

Across the river there was what looked like a suspension bridge. It was actually a cantilever, carrying a wooden flume raised on piles. Flumes and irrigation channels are very common in this part of the country, but I had not seen one conducted across a river before.

From Watak I saw a high mountain which I supposed was on the far side of the Salween, until I was told that we would have to cross it before reaching the river. The country changed and the valley narrowing to a deep gorge, the path began to climb the mountain on a long slant, still following the river, but rising steadily above it. The rocks were burning hot with the sun and it was tiring going.

When we reached the crest of the spur, I sat down in the aching sunlight to take bearings and study the new landscape. On the other side of the spur, 1500 feet below us, a river flowed from the south, as large as the river we had been following. The combined river, now considerable, flowed to the north through a deep cut in the mountains, and soon passed out of sight.

Beyond this new river the sheer mountain wall of grey limestone I had seen from Watak rose to a great height. A thin line was drawn high up across the mountainside. 'That's the road to the Salween,' one of the coolies said. What a road!

We descended a steep zigzag path to the river up which we marched for a mile or two. The path wormed between the river and the overhanging cliff. Sometimes it was squeezed into the river itself. At last we crossed it by a wooden bridge and came to Zigar, a small village where we changed our transport.

We could not follow the Tsa-chu down the gorge, and so we had to cross the mountain separating us from the Salween. We climbed steeply hour after hour, the path following the new river to the confluence, then crossing a spur, and following the river course again.

But whereas the river descended perhaps 200 feet a mile, we ascended 2000 feet a mile, and by the time we got back to the main valley we were 4000 feet above the river, which had dropped from sight. All we saw was its gorge, a bare slash in the measureless mountains rising all round us.

Towards dusk, reaching another crest, we saw a spot of green set in the bare rock, the village of Dza. The sun, which had long set in the deep valleys, caught the weathered mountain tops, and inflamed them. So high were those fluted spires that long after the world was in darkness below they continued to reflect the last stains of sunset.

We descended to Dza for the night. The temperature only fell to 57° and before dawn the flies made sleep impossible. It was like going down to inferno. The air smelt sour.

One of the most interesting and beautiful plants I discovered on this journey was a Leguminous shrub. At first I found it in ripe fruit only, and not recognizing its Lupin-like leaves and flat pods, I gathered seeds. It was abundant, though scattered, growing from 2 to 3 feet high. Presently I found a shrub beginning to flower—Laburnum yellow. I thought it might be a Sophora, though neither its fruit nor leaves were quite consistent with that. Still I could think of nothing else to call it. Above Zigar, where a stream welled from the slope, this shrub grew in quantity and I could guess from the abundance of its pods, the profusion of its bloom.

It was evidently a late flowering shrub—probably September or October, and that is always valuable in England. I was less certain of its hardiness. The wet atmosphere of England was against its thriving there. It would certainly grow well along the Mediterranean. But plants are so adaptable that I could imagine it growing happily in our eastern counties.

From Dza we climbed higher and higher, the view opening out behind us. We left the belt of xerophytic thorn scrub for thickets of shrubs and alpine flora, and passed from there to groves of Fir trees and Rhododendrons once more. It showed how high we had climbed above the gorge below; we were up at 15,000 feet again.

From here we saw the Salween, 6000 feet below, a great river, but running so muddy that it was hard to distinguish it from the biscuit coloured rocks. Yet the view was not as extensive as I had expected. I could see up and down the valley for some miles. But elsewhere, especially to the east, the view was restricted. I realized why; I was still on the plateau. The Salween did not flow in a gorge, strictly speaking. It cut through no great mountain axis here. All it had done was to plough this enormous valley in the plateau.

I looked round for snow peaks. But there were none, except behind me, where the range I had crossed stretched massive across the skyline for some fifty miles. Though it was sixty miles away, the visibility was so fine after the recent storm that I could see clearly the glaciers that fluted the dark rock. But there was no sign of other snow peaks anywhere in sight. My evidence was negative and disappointing.

We descended towards the Salween down a long slant, and reaching Puti, a four-housed village, 2000 feet above the river, at about five o'clock, we rested for the night.

CHAPTER XIII

THE RIDDLE OF THE SALWEEN

THE day had been exhaustingly hot, and the night also was close and oppressive. But the air was very dry; it was this extreme drought which accounted for the very different flora. It was a xerophytic flora, in appearance something like that of the Mediterranean coast. Woody plants, indeed, were not lacking; but they rather reminded me of Japanese dwarf trees. Many of them had, as you might say, a trunk, bearing a compressed crown, packed with flowers and leaves. Yet the whole plant would be only eight or ten inches high. Such was a charming Caryopteris, the stocky trunk immediately spraying up into a dome of twigs bearing innumerable tiny leaves and pale flowers. Other undershrubs were *Sophora viciifolia*, the dazzling blue flowered *Ceratostigma Griffithii*, a yellow-flowered Wikstroemia, Mentha, Buddleia, and *Rosa sericea*. The majority of them were either thorny or brusquely twiggy; which must have made them unappetizing even to goats. Yet goats picked up some sort of a living from them.

Plastered against the rocks were thousands of flat leafy rosettes. Several xerophytes assume this habit; the rosette curls inwards after the manner of a frightened sea anemone if water is too long withheld. *Didissandra lanuginosa*, with flowers like a pursed up Ramondia, and *Selaginella involvens* were the principal rosette plants.

Crops of barley and maize are raised; there are also many fruit trees, peach, pear, apple, walnut, mulberry; so that a village, its terraced fields green with corn, shaded by trees, is a real oasis in this rocky wilderness. Moreover at Puti *Clematis connata*, white Jasmine, and an Indigofera with long drooping racemes, like a purple Laburnum, still further beautified the terrace walls. There was also a Hawthorn with remarkably large leaves; I regretted that I saw this neither in flower nor in ripe fruit.

Even on the plateau, directly above Puti, the flora is xerophytic rather than alpine, although the altitude is not far short of 15,000 feet. But so dry is the country to the north of the Salween-Tsangpo divide that the snow-line is hardly less than 19,000 feet, and the alpine flora correspondingly elevated. There are flowering plants at 18,000 feet; at 17,000 feet they occur in considerable variety. I was therefore not altogether surprised to find groves of Fir trees in sheltered hollows on the plateau. What did surprise me was to find two species of Rhododendron here, associated with the Fir trees. For the rock was limestone, and Rhododendrons do not normally grow on limestone; it is poison to them. I found only four species of Rhododendron north of Shugden Gompa: two of these, *R. vellereum* and the aromatic-leafed *R. tsarongense,* are found above 14,000 feet, wherever the cliffs are of limestone. They grew here, but had not flowered this year. The other two, *R. sigillatum* and *R. telmateium,* grow anywhere except on limestone, thus behaving normally.

After a night's uneasy sleep I descended to the river by a steep and crumbling path. For the last fifty feet the gravel cliffs were almost sheer, and I had to scramble down a torrent bed. But eventually I stood by the mysterious Salween. A line of huge boulders marked the narrow shore. The turbid yellowish water seethed as it spun along at ten or twelve miles an hour, but this appearance of boiling was occasioned rather by boulders in its bed than by serious falls. There were no big rapids, still less falls; the gradient though considerable for so great a river was steady. Comparing the Salween with the Tsangpo in the same latitude we find them very unlike each other. Just above the great gorge the Tsangpo is flowing at a height of 9600 feet, the Salween at 9200 feet. But whereas the Tsangpo is only 300 miles from the plain of Assam, the Salween is still 900 miles from the plain of Burma.[1] I got the impression that the Salween has been scouring this valley for an immensely long period; it was a much more finished article than the raw and rugged Tsangpo gorge. The latter is a modern work, in course of construction, or else it has been recently uplifted at its head.

For all its scorching rocks and sapless grey-green shrubs the Salween valley was not devoid of animal life. Lizards (*Japalura splendida*) sun themselves on the rocks, and when alarmed scurry up the dusty slopes with erect tail. They are ashen grey in colour, with a pale yellow stripe down each flank, and a similar spot like a candle flame under the throat. I also saw a frog (*Bujo viridis*) with a green or yellow stripe down the back. Grasshoppers revel in sunshine; but they also need succulent vegetation from which to imbibe juices. They abound in the Mishmi hills, and I was surprised to see so many here. Small butterflies hovered about the villages, where they could be sure of offal, and hence, of salt. We commonly associate these beautiful and delicate insects with honey and flowers. But salt is no less necessary to their existence, and this they obtain by scavenging in the village muck heap. There is no bait so effective as garbage for attracting butterflies. Beetles were not so common though hardly less interesting. Only a fat oil beetle, black with vermilion bands, was abundant, crawling on the ground. It was mating. Birds, however, were rare, nor did I see any mammals.

Altogether the fauna and flora of the arid regions of eastern Tibet would repay further investigation. These regions are isolated areas sunk in the plateau, and it is important to discover whence the species which now occupy them have been derived. So far as I know the relationship both of the fauna and flora is with China rather than with India, or even the Himalaya.[2]

From Puti we retraced our steps as far as Zigar on the Aju Chu. It was my intention to follow this tributary to its source, cross the snow range by the pass at its head, and so reach Shugden Gompa by a new route. Looking up the valley of the Aju Chu from a spur above Zigar, I observed a remarkable conjunction of white limestone and red sandstone. Broad bands of these brilliantly contrasted strata formed a colour scheme on either side of the valley, and indeed the rocks were bent into troughs and arches right across the country, the general direction of strike being from east to west. Whatever the age of these rocks— and they were probably of Devonian age—they had been thrown into a wavelike series of north and south folds. The inference is irresistible that this folding was done when the Himalaya were uplifted, and by the same agency. But whereas the rocks which form the Himalaya then lay at the bottom of the sea, and were slowly uplifted to their present commanding height, the rocks I was now observing, if of Devonian age, were, and had long been, dry land.

I spent an unpleasant night at Zigar in the headman's house, for my room though large and airy was full of the most voracious bugs. My bed was soon seething with these foul vermin and I got little sleep. The following night at the next village fleas took the place of bugs, and my slumber was again disturbed. After that it was a relief to sleep in my tent.

The lower course of the Aju Chu, like the Tsa Chu, is well populated. Villages, though small, are numerous, but cultivation does not extend far, and we reached the last village in a day's march from Zigar. Here we halted on August 6th, as it took time to collect men and supplies for the five days' journey to Shugden Gompa. These villages are not, as one would suppose, from the fact that they lie north of the great range, under any of the Nagong Dzongpöns, but under Sangachu Dzong. Thus they belong to Zayul. My name stood high with the Governor of Zayul, and Tsumbi was able to get transport quickly by telling the headmen that I was a friend of his.

On August 7th the heat of the last few days culminated in thunderstorms. At noon the sun was blazing in a cloudless sky. By three o'clock the air was quite chilly, and it was raining steadily. Above Yindru, the last village, the sedimentary rocks which had interested me so much ceased and were replaced by granite. We ascended through a narrow chasm, the cliffs harbouring all kinds of unexpected flowers. Already we were out of the arid region and back to the temperate flora. There were small trees here lining the torrent—Poplar, Cherry, Juniper, and presently *Picea lichiangensis* again. The most conspicuous flowers were the prickly blue Poppy, a smart little Saxifraga with shining yellow flowers and a carpet-forming Androsace, whose pink and white flowers well matched its silvery hoary leaves. But most surprising was it to see solid tuffets of woody stemmed Primula on the granite cliffs—crumbling leafy hassocks which must have clung there probing ever deeper into the living rock for twenty-five years or more. The shrubby Primulas grow very slowly, almost imperceptibly; a quarter of a century would not be too long for those plants to have reached their present size. The flowers were past, but I recognized the plant as *Primula Dubernardiana*.[3]

From the temperate belt to the alpine region was but a step. Had the weather been fine I should have enjoyed the cool air off the ice, after the desiccated fly-blown atmosphere of the Salween gorge. But it now set in positively wet, and when we camped high up, preparatory to crossing the range next day, the cold was unpleasant.

August 11th was a day to remember for its discomfort, though also as a day of achievement. The clouds were sweeping steadily up the valley when we started, and there was no hope of a reprieve. At the head of the main valley stood a group of snow peaks with glaciers stuck on them. We turned aside; a man told us that if we continued up the main valley and crossed the range, we should come to a village called

Ru. I was destined to hear much more about Ru later. It insinuated itself into every discussion as to our route; it swelled to gigantic importance; for a week it dominated our lives. But I never saw it. That Ru is a myth I firmly believe, but it is a myth which has to be reckoned with.

Approaching the pass was like entering a cold storage chamber. Glaciers draped the converging precipices. They clung to the rock sprawling awkwardly; their blunt splayed snouts never reached the valley. They resembled not so much rivers of ice as cascades of ice. One only was different; its snout was broken off in a short ice cliff, and immediately below the cliff lay a tiny jade-green lake. But though I counted more than a dozen small glaciers I did not see one really snow-covered peak. The glaciers were fed from small snow-beds tucked away between the rocky peaks, which rose dark and forbidding all round. They were too steep to hold the snow; but also they were not high enough—evidently the highest of them did not greatly exceed 20,000 feet.

By this time the rain was descending in sheets and we were wet through. But the rain soon turned to snow almost blinding us in its fury. From the head of the valley a sharp climb over boulders and rock ledges brought us to a razor-backed ridge connecting two peaks. A glacier flowed down either side of the ridge. This was the pass called Traki La, 18,487 feet. From the top we looked down a gloomy valley, flanked by small glaciers none of which reached the bottom.

I was enchanted with the alpine flowers which grew here. It surprised me to find so many flowering plants between 17,000 and 18,000 feet, although isolated plants are occasionally found even higher. Most conspicuous was *Primula macrocarpa,* now in full bloom. Its hearty leaves, from which sprang robust stems, clogged the sand spits deposited by newly-born streams. The flowers are mauve, clustered into a fine head, and faintly fragrant. In the streams which trickled across the ice-scoured floor at the heads of the valleys were a wealth of semi-aquatic plants, and these were perhaps the most interesting of all.

Two species of Cremanthodium deserve notice. One had brass-yellow flowers, the ray florets finely drawn out till the nodding head resembled a half-open parachute. The other, a coarser plant, displayed its brazen fringed suns on the more inaccessible screes, where the crash of leaping water down the gully, and the grinding of gravel, loosened by the rain, were pregnant with unseen danger.

A big garlic was in fruit, and colonies of *Cochlearia scapiflora* deeply rooted in the shifting sand. As high as we climbed patches of egg-yellow *Draba alpina* and rose-pink *Androsace Chamaejasme* encrusted the cold grey rocks; while here and there *Saussurea gossypifera* poked up its white cone like a long woolly gas-mantle. One of the commonest aquatics was a yellow-flowered Corydalis with finely-cut leaves.

The drier turf slopes from where the steep gravel fans flattened themselves to rest, to where the streams began to cut channels for themselves, were covered with quite another set of plants. Here grew a dwarf Primula, like *P. minor*, but the flowers were over. Polygonum, Potentilla, Anemone, Caltha, Aster, *Meconopsis horridula*, Ranunculus, and Dracocephalum, are a selection. The Dracocephalum deserves some description, and I subsequently collected seed of it. The rather small violet flowers are in close heads; its charm lies not in the flowers, but in the large leaf-shaped bracts which support them. These are of tissue paper translucency and consistency, delicately veined: they almost hide the flowers, and give quite a different appearance to the plant. It grows nine inches high. Abundant and varied were the mossy Saxifrages. Some have cadmium yellow, others orange flowers, due to orange spots spread over the yellow ground of the petals. The stamens usually match the petals, but some species have coral-red anthers. The Saxifrages growing in clumps on the bare earth fans form golden nuggets; and sometimes looking across the fans in lambent light these nuggets seem to liquefy and flow together in a molten stream of pure gold.

No plant I found here, however, interested me so much as a certain aquatic Lagotis. At first glance I took it for an orchid! Had this been my first sight of the plateau flora, I should certainly have been bewildered and amazed. I was well content to be enchanted. On this day, crossing the range, I found forty species of alpines at 17,000 feet and higher, many of them new to me.

Owing to the snow and rain, and to the sharp wind which assailed us on the pass, I was numb with cold by the time we had crossed the ridge. The snow had given place to rain once more, but it was only thawed snow. For some distance I trudged on down the valley; finally I mounted my pony. When late in the afternoon we reached a nomad encampment and halted, I could barely stand up, because all feeling had gone from my feet. But the herdsmen made us welcome and, having changed out of my wet clothes, I was soon comfortably installed in my tent drinking hot tea.

This camp was at 16,000 feet, and I decided to stay here a day in order to make a representative collection of high alpine plants. I found a new globe flower (Trollius) here; the corolla, though made of separate petals, persists long after the fruits have begun to ripen, and turns a warm mahogany red. In this dead state it is even more beautiful than when the flowers are 'open'. A dwarf Monkshood, not ten inches high, but bearing two or three quite large glossy violet flowers, was just appearing. It is remarkable how late is the flora of this high country; there is a good reason for it. In the first place, before the end of June the alpine valleys are bitterly cold, the snow has only just begun to melt, and by its very melting it absorbs a large quantity of heat. Then there is no constant water supply for plants, all of which demand ample water throughout the growing season. Lastly, late autumn, after the summer rain and before the winter snow, is the fine weather season, and the topmost valleys are often warmer in September, at least by day, than at any other time. Consequently, July to September is the season of high alpines; many plants remain in flower till well on into October, when frost at last puts an end to them.

There were two nomad tents here, each with half a dozen inmates, men and women. They came from Yindru, and looked after herds of yak, sheep and goats numbering hundreds. The nomads' tents, which are black in colour, are made of coarse sackcloth, which looks as though it would let in the rain; but except along the centre line, where the two halves meet and the smoke escapes, they are rainproof. Three centre poles support the tent which is tied and guyed by a multitude of ropes sewn along the edges of the cloth. The earth is the floor. Inside towards one end is a fire, with the altar beyond, and on the altar are one or two images and butter lamps. Large circular iron basins for boiling milk stand by—or on—the fire; and alongside are hide bags full of *tsamba*, wooden churns, and a litter of ropes, saddles, nose-rings, and other impedimenta. The herdsmen do not have the tent quite to themselves. In addition to the dogs, there are usually a few baby yak, and perhaps a tame pig or a sick lamb or kid. It is a crêche and a nursing home, too.

During the night I heard the rain pattering on my tent, but towards dawn it ceased, and an unearthly silence followed. I grew so cold that I woke up. It was still dark; the whole valley lay under a quilt of snow.

Even the noise of the torrent was muffled. The herdsmen asked me to stay another day, but as there seemed little prospect of fine weather, I declined and we broke camp. It was cold work packing up in the snow, but eventually we got off.

All day we marched through the sodden scrub by the ever-swelling torrent. We passed two more nomad encampments, halting at the second for a drink of scalding milk. The coolies pressed me to halt for the night, saying that they would start very early the next morning and reach Shugden Gompa the same evening; but I saw no advantage in that, and we went on. No doubt it was an unlucky day, and I ought not to have travelled. But who does not like to dare fate! Then we came to a fourth encampment where we had intended to halt for the night. Unfortunately the tents were on the other side of the river, and our ponies wisely refused to face the broad swift stream, though I watched two yak from the herd on the other side plough their way ponderously across to us, curious to see who the intruders were. The water was three feet deep, and the current strong. From this point the going was bad, the narrow valley being flooded. Sometimes we paddled in the brimming stream, sometimes we squelched through bogs, and sometimes we pushed through a barrage of shrubs. At last wet through, covered with mud from head to foot, tired, cold, and extremely hungry, we halted for the night. There was a patch of solid ground under a cliff. So we were able to pitch the tents. I knew we were not far from Shugden Gompa, because I could see the welcome tip of Dorje Tsengen framed between the sides of the valley. It was now certain that this torrent was the Tsengu chu which flows into the lake just below the monastery; and so it proved.

Next day an hour's march brought us out on to Bütang, and so to the lake. The last two miles down the Tsengu chu was through a gorge the cliffs of which were sparsely covered with Fir trees. Men were busy cutting poles for the building trade.

CHAPTER XIV

NINGRI TANGOR

My first question on arrival at the dzong was: 'Have any letters come for me?'

The Dzongpön shook his head; no letters or news of any kind. It was now over three months since I had had a letter from home, and my latest news was early March; much might have happened since then. But if I had no news, at all events I could give free rein to my imagination. The local people could not even do that; their imagination had no scope. I became cynical and decided that news was of no real value to anyone; or at any rate that what we usually called 'news' wasn't. The only real 'news' had no 'news-value'. What is popularly called 'news' is a more or less exciting romance—founded on fact.

When I first came to Shugden Gompa, the jingle of mule bells, or the barking of the monastery dogs used to bring me out on to the roof to see if a caravan from Lhasa or from Chamdo[1] had arrived, the caravan leader perhaps entrusted with my mail. But no caravan came; it was nothing—a horseman from a neighbouring village, no more. I relapsed into indifference. I would be like the people amongst whom I lived. 'News' meant nothing to them. The most momentous decisions, the most grave utterances, the most profound meditations by the wise men of the West were, to these wiser men of the East, less than the dust. It was a thought to chasten pride.

A good deal of rain had fallen; evidently August is a wet month. A large pool of Guinness-coloured liquor—one could hardly call it water—occupied the yard where they milked the yak.

In the course of the next week, however, several events happened, and in the intervals of plant hunting I found much to do. First of all I must have a pair of Tibetan boots made for the winter. There are cobblers in Tibet—usually of Chinese extraction, but there was not a cobbler at Shugden Gompa. The nearest was said to live at Sangachu Dzong. So I sent a letter to the Governor of Zayul, who had moved from Rima to his summer capital at Sangachu Dzong, requesting him to send the cobbler to me. The letter was taken by special messenger, who got back with a reply on the fourth morning. The Governor informed me by letter that the cobbler had gone to Draya Gompa, three days journey to the south. He had, however, commanded him to report to me at Shugden as soon as possible. He also sent me a present of two pieces of soft leather, one black, the other scarlet, from which the boots were to be made.

At this time I was living a life, not indeed of ease, but almost of luxury, with green peas and turnips from the fields, and as much milk and butter as I wanted; but the peas did not last long. I had some dried raisins in tins; and Tsumbi, who was now my cook, could make an excellent curry.

One of the first excursions I made after my return was to the summit of Ningri Tangor, to search again for the rose Primula. It had now become a point of honour with me to find that plant. Tashi and I rode up the ridge on ponies; the last part of the ascent was very steep, but the sturdy little animals made light of it. Beneath the Rhododendron bushes I found several plants of a gorgeous Gentian. It grew quite erect but was not above nine inches high, the leafy stem ending in a short spire of cobalt blue, narrow funnel-shaped flowers; they stood upright, actually in the axils of the topmost leaves; but the leaves grew so close together that the flowers formed a solid spearhead. This was G. *trichotoma*, a plant which, raised from seed collected by me in China, in 1921, hovers between cultivation and extermination. It requires all the skill of the experienced cultivator to maintain it, but its rare beauty rewards the successful grower.

Along the crest of Ningri Tangor was a mile-long shallow depression, floored with grass and alpine flowers. Most of this depression was dry, but at one end was a basin lined with greener grass, though I never saw any water in it. Moister it certainly was; the flowers inside were always different from those beyond its rim. At this season the whole basin was covered with a mauve carpet of dwarf Asters; there were millions of them. I say dwarf in preference to pigmy, because the plants *were* out of proportion and not merely miniature. The stem did not exceed two inches in height, yet the flowers were that much across. Against the distant snow-glazed ranges under a turquoise sky, the green alp suddenly changed to a mauve film of flowers; the contrast was superb. On the grass slopes, another Gentian, very different from G. *trichotoma*, was opening its large goblet-shaped trumpets. These were borne singly

THE LAKE, SHUGDEN GOMPA

at the ends of radiating shoots, which half-rose to carry them upright but failed, so that they lay almost prone. The trumpets were large with frilled mouth, purplish in colour and beautifully spotted. I recognized *Gentiana Georgei*, a rock plant which ought to be but is not in cultivation. But the large rosettes of this fine Gentian were almost swamped beneath the deep blue waves of *Gentiana sino-ornata*, not yet in the spring-tide of its glory. *G. Georgei* and *G. sino-ornata* grew together side by side; yet they never hybridized—or if they did, no hybrids survived. But *G. trichotoma* grew aloof. In September, when the dwarf Aster was over, and Gentians swept in irresistible and incredible blue seas over the windy crest of Ningri Tangor, a Cambridge blue form of *G. sino-ornata* filled the basin; the navy blue form stayed outside.

So we returned to the monastery rich in Gentians. It had been a successful day; yet there had been a failure to mar my contentment. I had seen no sign of the rose Primula.

On August 19th a visitor arrived from Sangachu Dzong. This was the junior officer whom we had previously met at Rima. He brought me presents from the Governor—bricks of Chinese tea and of sugar, and packets of Chinese macaroni. Tsumbi brought him in to see me on my return from an excursion, and I gave him tea. In his blue *chupa*, long black leather boots, and Homburg hat, he looked rather more impressive, despite his youth. He was doing a little private trading, government business being slack.

I spent the evening writing letters; the young officer had brought me a message from the Governor that he was shortly going to Chamdo, and would take letters for me, to be forwarded from there to Lhasa by the official post, and sent thence to India. These letters, eight in number, left Shugden Gompa on August 20th and reached England in December. On the same day I finished reading *Bleak House*. I had with me only two novels, the other being *Pickwick*, besides Boswell's *Life of Johnson*, and the complete works of Shakespeare; and I was sad at finishing *Bleak House*, which had whiled away many an intolerable hour of waiting for transport. I might almost say that for weeks I had lived with many of the characters—turning to them whenever I felt lonely or depressed or disappointed, and never failing to benefit from their presence.

My next objective was Sangachu Dzong, but I intended to postpone my visit there until September. The only other direction in which I could explore was north-westwards down the Nagong Chu, and as pack animals could only go two days' journey, that was as far as I intended to go now. While waiting to start for Sangachu Dzong I therefore decided to follow the Nagong river.

This time we followed the shore of the lake northwards along the western foot of Ningri Tangor. The path climbs high up the cliff, descending again at the north end of the lake. Near the lakeside at a village called Ruowa we halted for the night. The Poyü La torrent from the north has thrown out a delta, again constricting the lake, which turns sharply westwards, continuing for some miles. As at Shugden there is excellent pasture. But Ruowa has another and more important industry than either pastoral or agricultural. After marching for three hours next day we reached the end of the lake, whence the Nagong river issues to plunge into a thickly-forested gorge. Indeed all along the lakeside from Ruowa westwards trees grow. I observed several timber rafts on the lake, and at Ruowa poles were being loaded on to yaks. Each beast carries or rather drags two on either side, the ends trailing on the ground behind them.

Before we reached the foot of the lake we passed two more villages, the houses built of timber with high-pitched roofs, more like the houses of Zayul but taller; there was also a little monastery on an island but it was uninhabited. There was no doubt as to where the lake ended, for the silence was suddenly broken by a roar as the turbulent river rushed into a rock-strewn gorge. Moreover the rock changed suddenly from limestone to granite. Continuing through the gorge, along a shelf of water-worn cliffs, we camped in a grassy clearing beneath magnificent Fir trees, and here I remained all the next day. A very rickety cantilever bridge spanned the river; it canted dangerously. From this point there is only a difficult footpath through the gorge. It is said to take ten days to reach Showa, the capital of Pomé; but there are several Poba villages, the first at a distance of not more than two days' march. The river here is called the Nagong Chu, but there can be no doubt that it is the same river as the one elsewhere called the Po-Tsangpo, which joins the Kongbo Tsangpo at Gompo Ne. There is an unexplored stretch between my camp on the Nagong Chu and Showa, which would be interesting to a botanist. I could see only a short distance down the gorge on account of the forest.

Of the high peaks previously observed to the north-west I could now see nothing; the only snow visible was up a wide valley to the south, in the direction of the Lhagu peaks.

I would have liked to follow the Nagong river further, but was unable to do so as I had promised the Dzongpön that I would not go into Poba territory. It seems that the Tibetan authorities are a little shy of their independent neighbours, who have the reputation of being wild and lawless. The word Poba means no more than an inhabitant of Po; the two divisions of Po being Pomé (Upper Po), and Poyul (the country of the Po). But Po or Bo is the same as Bod, the Tibetan name for Tibet, which suggests that the Po country is the original home of the Tibetan race. The Pobas are short and stocky, with Mongolian features. They wear their hair long, not plaited into a queue, but they dress like other Tibetans.

Occasionally Pobas are said to visit Shugden Gompa, but none came while I was there, and they are not encouraged to do so. Nor did I meet anyone who had been down the valley of the Nagong Chu. Below this point the gorge rapidly becomes moister, as indicated by the appearance of forest consisting principally of *Picea likiangensis*, with two species of Birch forming a lower growth. I noticed an Abies here also. In the forest I saw a woodpecker, the first I had seen north of the Ata Kang La: and in the evening many daddy-long-legs (*Tipulidae*) came into my tent; all indications of a clammier and more thickly wooded country. The altitude of the river here was 12,643 feet, and the altitude of Showa is 8520, a difference of 4123 feet. This gives an average fall of 41 feet per mile. The gradient, however, is not likely to be uniform, and is probably much less than this towards Showa. There may be very big rapids lower down; and indeed the Nagong Chu was already a monstrous torrent, roaring between sheer granite cliffs about twenty-five yards apart.

After staying a day at this camp to collect plants and make observations I returned to Shugden Gompa. Approaching the monastery along the lake shore I saw the clouds towering straight up over the distant snow range like smoke from some gigantic fire. It was as though they were held back by an invisible force. Suddenly everything seemed to give way and they came romping over the barrier. Now they lost their silver lining and firm outline. The wind howled and whistled over the dead earth, tongues of lightning flickered in and out, and snow began to fall. But the storm circled round and soon disappeared again.

Great activity prevailed in the village. The short autumn was upon us with the long bitter winter close on its heels. The crops were almost ripe. Women were coming into the monastery and the dzong with bundles of firewood. The monks were hard at their studies, reading or reciting aloud, as in the village schools at home years ago. Term lasted for a month and five days, after which the student monks might get a few days off. Ignorant of the rules and regulations, I again requested the abbot to give a day's holiday to the lama who had brought me the rose Primula. But the answer came back, he was busy at his studies, and I must wait till the vacation. Once more Tashi and I rode up Ningri Tangor. It was a lovely day. Countless thousands of Gentians sparkled on the crisp turf—*Gentiana sino-ornata* and *G. Georgei* with flowers an inch and a half across. In one spot there were forms of *G. sino-ornata* with white flowers, and others of pastel shades. Most beautiful of all was a Cambridge blue form which now occupied the damp basin previously occupied by the Aster; and I discovered that by streams, and in marshy ground generally, where for example *Primula tibetica* flourished, it was always the pale blue form of *G. sino-ornata* which occurred. Though I spent the greater part of the day on Ningri Tangor, and searched all the most likely places for the rose Primula, I still could find no trace of it. I felt completely baffled; unless the monk could show it to me, I had little hope of ever finding it now. But could the monk find it? Six weeks had already elapsed since that July day when he had brought it to me. The flowers were long since over, and it was hardly likely that he would recognize one particular flower out of a dozen from my description or recollect where he found it. I began to think I had seen the last of that Primula, and I blamed myself severely for not having kept the rather jaded specimen he had presented to me.

When I returned to the dzong, rather disconsolate, I found visitors from Sangachu Dzong waiting to see me. Chimi had returned bringing with him the cobbler. The third member of the party was Kele, Tsumbi's brother from Chamdo, who had come to seek service with me.

Chimi said he had accompanied Ronald and Brooks-Carrington as far as Giwang. There he had left them, and apparently become a gentleman of leisure for a time. He had then wandered round to Sangachu Dzong and feeling the pinch of unemployment had come on to me at Shugden Gompa.

Kele kept a shop in Chamdo. Having heard from Tsumbi that we were bound for Shugden Gompa he had made the long journey—it had taken him twenty days' hard travelling—partly to see his brother and partly to seek service with me. He was a short stocky little man, very bandy-legged, so that he walked with a waddle. He wore a Homburg hat with a piece of Indian silver braid round it, and Chinese boots. A large silver ear-ring with a turquoise set in it dangled from his left ear. Otherwise his dress was Tibetan. He spoke no Hindustani, and his Tibetan was incomprehensible to me. But he was a willing knave, and seemed anxious to serve me and to learn; so after a trial I took him on permanently.

The abbot invited me to lunch again on September 2nd, and I went because I wished to approach him once more about letting off the monk to show me the Primula. They gave me a much better meal this time, some mutton—or more probably goat—fried in oil, being delicious. There were also sausages a foot long, but what they were made of I never discovered. Several monks—cooks and others—stood around to assist me on this occasion; it seemed to be an informal party. One monk, seeing me in difficulties trying to cut up the meat on a tin plate with a clasp knife and a pair of chopsticks, wiped his not very clean hands on his rather dirty gown, picked up the lump of meat off my plate with his soiled fingers and cut off morsels which he offered me. I ate them. After six months in Tibet the edge has worn off one's fastidiousness.

Again I requested the abbot to give the monk a holiday: this time he agreed to do so on condition that I sent to the monastery the sum of three *tankas* (about ninepence) and a scarf of greeting; the money was to provide a substitute at prayers! The fiscal arrangements having been satisfactorily carried out, early next morning the monk appeared, in high spirits at getting off schools, and we all rode up Ningri Tangor together. My guide was dressed in his red robe and long red boots and wore a fur cap. Approaching the highest peak, the monk outdistanced us and disappeared over the ridge. Tashi and I followed. Arrived at the summit I could see nothing of our guide, and fearful lest he should uproot the Primula, I called to Tashi to find him quickly, as I would not countenance the unnecessary destruction of wild flowers even in Tibet. His anxiety to prove that he knew exactly the plant I meant, and where it grew, was both comic and touching. At last the monk reappeared, ascending the cliff which faced the lake. He held a plant in his hand. I looked at it, and without a doubt it was the lost rose Primula in half-ripe fruit! After all it grew on the scree which I had dismissed as impossible, and yet explored to make sure! I had been within a dozen yards of the plant on my second visit to Ningri Tangor: had it been in flower I could not have failed to see it. Yet deliberately not expecting to see it I had passed it over. I felt conscious of my folly. I now descended the slope, and found perhaps a couple of dozen plants scattered about—it was not abundant—with *Paraquilegia microphylla* and a few shrubs. It was certainly the last place I would have searched for a 'Nivalis' Primula; hot, dry, and windy in summer, but cold enough in winter, with plenty of snow. I was very pleased with the monk. Who dare say now that the Tibetans are not born botanists when they can recognize plants as accurately as this man did?

Never again shall I lightly cast aside a flower brought me by a native, on the easy assumption that I can collect it for myself at leisure. Four days I spent searching Ningri Tangor for the rose Primula; I did not find it, nor did I meet with it again elsewhere, until October, when I discovered a few plants some distance to the north of Shugden Gompa—not in flower of course. What struck me particularly was the accuracy of the monk's observation. He *said* he found it on the summit of Ningri Tangor, and he *did* find it there; only my conviction that it could not grow on the hot dry shale scree had prevented my finding it for myself.

We returned to the monastery and the monk went back to his books. Before we left Shugden Gompa I collected abundant seed of the elusive Primula.

CHAPTER XV

THE DESERTED FORT

I WAS now ready to go to Sangachu Dzong. But first I must have some boots, mine were quite worn out. So the cobbler got to work and made me a pair of long thick-soled Tibetan boots. I told him to make them large so that I could wear two pairs of thick socks in the winter and not feel cramped. He therefore drew a rough outline of my foot, and then made the boots larger—so large that I could almost have got both feet into one. In these remarkable boots, padded with straw, I slopped about for three months. At least they were warm. I watched him at work with some interest. He used two needles at once, and strong twine, boring holes in the leather with a sort of bradawl. Every time he drew a thread through the leather he spat on the work to cement it.

On September 6th we started for Sangachu Dzong. The coolies, who had received two days' notice before, arrived late, and the Dzongpön, aware of my impatience to start, was furious. A woman who was responsible for supplying two ponies was the last to arrive, and she received the full brunt of the Dzongpön's displeasure. Having rated her soundly, he beat her over the head with a stick until Tsumbi told him to desist. He was really angry, his face distorted with rage, and was breathing hard as a result of this manual labour. As for the woman she simply protected her head and face with her arms, and took her castigation as a matter of course. She seemed neither surprised nor hurt.

A remarkable feature of Tibetan administration is the absence of force behind the government. Neither in Zayul nor in Nagong did I see a single soldier or policeman. The Governor of Zayul, being a person of importance, did not need to inflict corporal punishment himself; his servants did it for him. The Dzongpön of Shugden, having only one senile servant, was his own executioner. Beatings are a common form of punishment.

Travelling south-eastwards we camped in the mile-wide grassy valley of the Yatsa Chu, between rugged ranges of mountains, seamed with hanging glaciers. The Yatsa Chu itself is invisible from a distance of fifty yards; it has cut a narrow trench a hundred feet deep in the gravel floor of the valley and buried itself out of sight.

From here to the pass over the watershed we hardly climbed a thousand feet in eight miles, so easy was the gradient; in fact it was difficult to tell we were on the watershed, which is remarkably flat for several miles. It was splendid grazing country, a well watered undulating plain flanked by icy mountains. Millions of alpine flowers frisked and shivered in the gusty wind from the snows: and indeed the higher we climbed the more flowers we saw. I was pleased to see *Gentiana trichotoma* covering acres of gravelly soil. It grows much more scattered than *G. sino-ornata* and its allies, but there were thousands of plants here, and the flowers varied from pale blue or almost white to a deep sea blue.

The fact that we had crossed the watershed was soon made evident. Two streams, one from either side of the valley, united, and flowed southwards. It was not possible to follow this torrent any distance; it soon plunged into an impassable gorge, amongst snow peaks and glaciers, east of the Ata Kang La. The track thereupon ascended a steep spur to a pass called the Zo La,[1] whence we could see far down the valley to the south, and easily make out the white monastery of Sangachu Dzong, perched on a lofty rock. No snow was visible in this direction. Though it was early autumn by the calendar the alpine flowers were now at their best: even on the stony summit of the Zo La there grew clumps of a beautiful golden Anemone. This formed mounds of yellow buttons set in green foliage all up the scree. Another notable plant here was *Gentiana Wardii* whose leaf mats sparkled with bright blue glass bubbles. Down the steep slope on the other side of the pass, the sulphur yellow bells of *Primula sikkimensis* chimed with the mahogany globes of Trollius, which was passing out of bloom. There were many other alpine flowers,[2] and after we had descended a thousand feet, and camped close to the first trees, I noticed several dwarf Rhododendrons which I had seen in bloom on the Ata Kang La in June. The change in the type of vegetation, after we had crossed the pass, was very marked—not in the alpine region, but lower down. Here forest reappeared. The trees were neither so large nor so varied as in the Rong Tö valley, but at least they were trees.

On the following day we continued the descent into the wooded valley. We soon reached the junction of the stream from the Zo La with the main torrent, called the Tsiling Chu, which, as we have seen, rises on the watershed north of that pass. Looking back up the gorge from which it emerges, I again caught sight of glaciers and snow peaks. We were now following the headwater stream of the main Zayul Chu, the eastern of the two branches of the Lohit which unite just above Rima; in fact we were on the main road from Shugden Gompa to Sangachu Dzong and Rima. The valley is cultivated for several miles, with houses now on the left bank of the river, now on the right. Finally we crossed the Tsiling Chu by a wooden bridge, and passing by a considerable expanse of cornland, where monks and villagers intermingled were reaping the corn, we climbed a steep path to the monastery and fort of Sangachu Dzong.

The monastery straggles along the ridge, high above the river, and except that it has a wooden tiled roof, is similar to that at Shugden. There were very few monks in residence. Some were at work on the estates, some were away. All told, the monastery might support a hundred. The dzong to which I repaired stood next to the monastery. It was a substantial building, entered through great wooden doors which led into a dark courtyard and stables. A flight of wooden stairs, each stair very steep and narrow, connected the courtyard with a gallery, with numerous rooms opening off it. The flat roof was reached from one corner of the gallery by a ladder.

Except for a caretaker, the dzong was completely deserted, the Governor of Zayul having departed for Chamdo. The caretaker gave me a large airy room, and my servants established themselves opposite me. I was glad to be alone; but before the third day I had reason to regret that the dzong was empty. There are four villages down by the river below the fort, one downstream, and three upstream. There are also four bridges over the river. Each village is on duty for three days at a time. The villagers have to supply firewood and perform any services the Governor may require. It happened to be the turn of the village farthest away, and it was some time before I could get hold of the headman and tell him to keep in touch with me. I also told him that I would remain here two days, and that on the third day I wanted transport to take me back to Shugden Gompa.

On the afternoon of our arrival I explored the dzong and the monastery, which commands a good view up and down the deep valley of the Zayul Chu. Northwards the Zo La which we had crossed

A MENDICANT MONK

THE ABBOT OF SHUGDEN GOMPA

is conspicuous: to the west of it is the group of snow peaks at the foot of which the source stream flows in a gorge. Eastwards the view is shut out by a rocky range, somewhere beyond which flows the Salween. Immediately below the fortress a wild torrent rushes through a rocky gorge, and looking up this gorge I saw another group of snow peaks and glaciers, evidently close to the Ata Kang La; only now I was seeing them from the east, or more properly the north-east side. Evidently I was not very far from Ata and the Rong Tö valley; and I now learnt that there was a path up this valley to Ata via Suku, though it would take me at least a week to get there. The path is only fit for coolies and is little used.

It was this torrent which supplied both monastery and fort with water. We were some three hundred feet above the torrent; but the water was brought to a point fifty feet below the fort by means of a flume, which drew it off the torrent two miles higher up and led it round the side of the ridge in a swift stream. On the following day I walked along the embanked flume, finding the steep sheltered face of this limestone mountain to be thickly overgrown with shrubs and small trees.[3] But the most interesting plants here were a pale Adenophora, like a many-flowered Harebell, a form of *Lilium Wardii* and the pink Androsace of the Salween. I was able to secure ripe seed of the Androsace.

The eastern Zayul valley is not so well wooded as the western Rong Tö. However, *Pinus Khasia* grows here, and higher up are Picea and Abies. A fine looking bush Rhododendron related to *R. Thomsonii* grows at 12,000 to 13,000 feet, together with a very aromatic leafed shrub related to *R. anthopogon*. Whether it had pink flowers like a similar one collected on the Ata Kang La, or white like *R. tsarongense*, I do not know. It formed a fine bushy shrub four feet high and as much through, and had been a mass of flower, but the plants I saw were not less than twenty years old and may have been older. A third species was *R. lepidotum*. What surprised me was that here were limestone cliffs, with Rhododendrons growing not merely happily but luxuriantly upon them. Other shrubs were species of Rose, Viburnum, Dogwood, Honeysuckle, Deutzia, Birch, Willow, Mock Orange (Philadelphus), and Litsaea.

I was interested to find *Lilium Wardii* here, at an altitude of 12,000 feet; it is known as a plant of comparatively warm valleys. The plants were small, and had borne one or two flowers only.

THE DESERTED FORT

Many ground orchids grew on the grassy limestone rubble slopes, especially a large leafed Cypripedium, which might have been *C. luteum* or *C. tibeticum*. Here, too, I got seed of *Onosma Hookeri*, a plant which chose the most inaccessible crags.

That night after dark a weird blaring sound came from the rocks above the monastery. It was the monks blowing trumpets. The deep notes of the long trumpet reverberated from cliff to cliff, and every now and again were interrupted by the shriller notes of the short trumpets. There was something uncanny about the sound coming out of the night like that in the fitful glow of a wood fire. Did the trumpets utter a warning? It almost seemed as though they did.

CHAPTER XVI

THE FIGHT IN THE DZONG

Next day I cancelled the order for coolies, having decided to spend another day plant-hunting on the cliffs. I took Tashi, Kele, and Chimi with me; only Tsumbi stayed behind in the dzong. We spent an interesting day on the limestone crags, and it was late in the afternoon when we got back. Tsumbi did not appear when I called him, but after some delay he came, and I found fault with him. He seemed to comprehend nothing, and I dismissed him to the kitchen, telling him to hurry up with my dinner. But dinner was a long way off, and suddenly a row started in the gallery just outside my door. I went out. Tsumbi, already deep in my black books, had hold of Tashi and appeared to be threatening him. I went up to them and pulled Tsumbi roughly off. 'Stop it!' I said angrily.

He turned on me and was about to be insolent when I cut him short. I saw now that he was drunk. 'Be careful what you say, Tsumbi.' I seized his thumbs, ju-jitsu fashion; in a moment he was on the floor. He uttered no sound when I threatened to break both his thumbs if he showed the slightest sign of violence. But he was quiet enough and I let him get up.

'Go back to your room,' I said. 'I will talk to you afterwards.'

But Tsumbi was past caring about anything. I had hurt him, and he was as angry as I was. A few minutes later there was an explosion of oaths in the kitchen as Tsumbi cursed everyone in turn. Still thinking to quell the disturbance instead of letting the atmosphere cool down a little of its own accord—which might have been the wiser course—I went into the kitchen determined if necessary to have Tsumbi shut up for the night. Previous outbreaks had revealed him as noisy, quarrelsome and truculent when drunk. Tsumbi, wild-eyed and dishevelled, was sitting huddled up on a stool, shouting at everybody; his brother was on his knees before him, his head on his lap, weeping audibly; Chimi and the caretaker were cowering in a corner; only Tashi was trying to restore order. He spoke soothing words, and when I appeared, tried to drag

Tsumbi to his feet. But Tsumbi would have none of it. He was far too drunk to care for anybody or anything. With misty swimming eyes he gazed at me and shouted unintelligible things.

'Tashi,' I said, 'we must get him to bed.'

'I cannot do anything with him, *sahib*. Only Pinzo can control him when he's drunk.'

Then Tsumbi got up and in menacing mood advanced upon me, though handicapped by the weeping Kele: I felt sorry for Kele somehow.

'Tsumbi, get to bed immediately, or I shall hammer you.'

But Tsumbi was deaf to threats or to advice. It was clear that there would be trouble. The miserable Chimi chose this moment to start whimpering, and slunk from the room; but Tashi stood his ground like a man. Then Tsumbi burst out, speaking curiously enough in English: 'All ri, I know you,' he said shaking his fist at me. 'All ri.'

I gave him one more chance, then as he advanced I hit him between the eyes. The blow checked, but did not stop him; recovering himself, he aimed a savage but quite wild blow at me, grunting furiously. Kele threw himself between us, clasping me round the knees. I absurdly imagined that he was trying to protect me, or at any rate to part us; but as Tsumbi came on again, he held me tight, and I realized that he had thrown his weight into the scale on behalf of his brother; he was drunk too. Then I hit Tsumbi hard, and he went down. At the same moment Kele threw me, and of course I fell straight into the arms of Tsumbi, who was now like a maniac. I saw murder in his eyes, but I grasped his thumbs again and he was helpless. Tashi now pulled Kele aside, and after a minute I sprang up.

'Get him out,' I panted. 'Hold him, tie him up. He's mad.' All this had happened in less time than it takes to tell, but I was so out of breath from the scuffle that I could only sit and pant. Tsumbi and Kele now retired to the caretaker's room at the other end of the gallery, and I took a padlock from one of my boxes and locked the door, and I hoped that the incident was closed. Half an hour later there came a loud knocking, and Tashi reported that the drunkards were battering down the door. I went to look; but though the blows reverberated through the empty dzong the lock was a stout one. I believed it would hold.

Nevertheless my position was unenviable. Tsumbi meant murder—if he got out—of that I felt certain. Nor was I reassured when Tashi asked me meaningly whether I had a revolver. Of course

I had no such thing. Chimi and the caretaker had fled. Tashi and I were alone in the fort with those two drunken maniacs trying to get at us, and not another man near. It was now about nine o'clock. I had not dined, nor did I want to. I told Tashi to barricade himself in the kitchen, and to call me immediately if anything happened. Then I selected a stout cudgel about three feet long which I placed at the head of my bed. Finally I barricaded my door and set the table on its side so that anyone coming in must blunder into it and the noise would wake me. I had no mind to be murdered in my sleep. By this time I felt thoroughly frightened, and a long silence from the far end of the gallery was even more fraught with unpleasant possibilities than the hammering had been. But barricaded in my room, with the cudgel close at hand, I felt more confident. I was determined to use my weapon without hesitation if Tsumbi entered my room in the night. When everything was ready, I placed a powerful electric torch under my pillow and got into bed. In spite of a long and exhausting day it was some time before I got to sleep.

Dawn was breaking when I awoke. There was silence everywhere. There had been no dramatic turn of events. I dismantled the barricade and Tashi brought me my tea. I dressed quickly.

Suddenly a loud banging on the far side of the dzong warned me that the assault was being renewed. Tashi rushed in: 'They are breaking open the door, *sahib.*' No need to tell me that! With a loud splintering of wood a panel came out; the lock had held, the thick wooden door had given way. The prisoners had loosened a post and used it as a battering ram. In a minute Kele stood before me, gesticulating and speaking rapidly, in a mixture of Tibetan and Hindustani. I spoke sharply to him, refusing to discuss anything until he was sober, but though I put a bold face on it, I was feeling very uneasy. Tashi and I had to face these two thugs alone. Moreover, it now appeared that I had shut them up with the alcoholic supply—crude Tibetan rice spirit, and the sudden cessation of the assault on the door the previous night was only due to the fact that they had made this happy discovery. They had spent the hours of imprisonment very comfortably in the caretaker's room, drinking more deeply!

Knowing that Tsumbi would be upon me in a minute I pushed Kele out of my room and walked along the gallery towards the broken door; I felt safer in the open. Before I was half way, a figure appeared coming slowly down the gallery towards me. It was some seconds before I recognized Tsumbi, and I waited while he slowly and deliberately approached. He was a dreadful sight. He had a black eye, swollen lip and a cut forehead; a smear of blood striped one side of his inflamed face. His clothes were torn and dusty, his thumbs which I had wrenched back were swollen and discoloured. His pigtail had come down and his tangled hair hung over his forehead. But the look in his bloodshot eyes was defiant, insolent, and alarming. I had purposely brought no weapon with me, and seeing the pitiable state Tsumbi was in I decided, while relaxing none of my vigilance, to give him a chance of making peace.

'Well, Tsumbi, are you going to behave yourself?'

'Why did you hit me, *sahib*? I have been shamed!'

'You have forgotten something. Are you a coolie that you try to hit me? I thought you were my sirdar. What will they say in Darjeeling when they hear you were drunk and fighting with your *sahib*?'

In a moment the truculent expression vanished from Tsumbi's face. Without warning he was on his knees, clasping me round the legs, bowing his head on my feet, and crying bitterly. 'Forgive me, *sahib*, forgive me! I have a wife in Darjeeling, and a little girl.'

So that was that. I felt an enormous relief surge over me. But Tsumbi was far too dangerous and I determined to rid myself, if possible, of this turbulent sirdar. I now raised Tashi's wages for standing by me so pluckily during the crisis, and discussed with him what was to be done. He agreed that Tsumbi when drunk was uncontrollable. On the other hand if I dismissed him, Kele would go, too, and I would be short handed just when I most needed help. After turning the thing over in my mind I decided that I would not instantly dismiss Tsumbi, but that he should accompany me back to Shugden Gompa.

The coolies, having heard of the riot the night before and doubtless having been warned by Tsumbi not to come as he was unfit to travel, stayed away. I spent a restless day in the fort; and Tsumbi, retreating again behind the splintered door, slept and drank alternately.

I sent for the headman, and told him I must have transport for the morrow. This was promised, and next morning the men turned up after breakfast. Tsumbi, still looking a horrible sight, came out of his lair and took charge. He had pulled himself together, bound a turban on his head, washed his face, and done what he could to tidy up.

We retraced our steps up the steep valley. At the last village I noticed masses of Paeony growing on the edges of the field, but there was not a fruit to be seen. It seemed as though all the flowers had been cut, probably to decorate the household shrines; Tibetans usually put flowers on their altars. But I doubt whether this Paeony was indigenous. More likely it was *P. Delavayi,* introduced from China. On the sunny side of the valley Poplar and scrub evergreen Oak replaced Birch; but higher up the valley we came to real forest with *Rhododendron Beesianum.* I found a Cherry in ripe fruit here, probably the almond scented one already collected in the Rong Tö valley.

We camped on the tree line that night and next day recrossed the Zo La. There were still lots of flowers fidgeting in the wind. It was a fine day, and the distance had cleared up. The tip of Dorje Tsengen was visible. Towards the south-east I saw two conspicuously high rocky peaks, and in the north-west a sharp limestone fang rose above lesser peaks. The snow range lay immediately to the west; and farther I began to get a insight into the geography. I had been surprised to find that the Salween did not cut across a range of mountains as I expected; but clearly it could not do so because the main mountain range lay to the west, its axis pointing roughly north-west to south-east. This range has two crest lines, separated by a plateau region, thirty or forty miles wide. On the southern crest line is situated the great snow peak of Chombö, and the peaks to the south-east which I now saw for the first time. This crest is crossed by the Ata Kang La. On the northern crest line are situated the high peaks around the Traki La. As for the Zo La, it is not on the crest line at all, but on a spur; and the Tsiling Chu, which rises behind the Zo La and flows south-eastwards, rises on top of the range. It has worked its passage along the strike of the rocks, parallel to and between the two crest lines. Similarly the Nagong Chu flows north-westwards parallel to the crest line; hence it, too, rises on top of the range, and not on its flank. But if that is so, both the Zayul Chu and the Nagong Chu must cross the crest line of the range somewhere. The Zayul Chu flows southwards and then makes a great bend to the west before resuming its southward course at Rima. It must therefore cross the range, which lies immediately to the west of it, above its confluence with the Rong Tö river; and it must cross it in a gorge. I felt convinced that where the Zayul Chu turns westwards in the neighbourhood of Draya Gompa, it is cutting its way through the south-west crest line of the great range. As for the Nagong river, that I knew flowed in a profound gorge below Showa. Lord Cawdor and I had seen it and travelled up it from the confluence of the Nagong river with the Kongbo Tsangpo in 1924.

We continued the march towards Shugden Gompa, but in the afternoon turned aside to reach Lhagu, following the Yatsa torrent to the west. That night we slept at Yatsa, near the southern end of the lake. The barley was still green and unreaped; this at an altitude of nearly 14,000 feet. We reached Lhagu next day, September 15th, and on the 16th went into camp by the lake I had observed in July, under the wall curtained by glaciers. We pitched the tents on a grassy knoll, where the turf was quick with the vivid blue of a trumpet Gentian. This was not *G. sino-ornata*; the trumpets were not narrow funnels, as in that species, and the gussets between the lobes which enable these flowers to close at night were almost as deep a blue as the lobes themselves. I called it the Lhagu Gentian, and I found later that it was the commonest species around the Ata Kang La, but that it did not extend northwards on to the plateau. The common plateau species is *G. sino-ornata*. The low autumn sun lit up the lake and the frozen cascades behind, and shone briefly on the grass till the crowded Gentian cups sparkled like cut glass.

We spent three days here, my time being divided between collecting flowers and collecting seeds. The mornings began with drizzling rain off the snow peaks; the afternoons ended in a golden glow of sunset. All day the wind roared and whistled amongst the glaciers, buffeting me as I scrambled up and down the alps to get fresh peeps at this or that glacier. There were no less than four lakes here, occupying deep rock basins, and all of them connected with the Lhagu lake, and thence with the Shugden lake. This chain of alpine fjord-lakes extends for a distance of twenty miles before the Nagong river finally leaves the Shugden lake to crash into its gorge. At this season I had a good view over the ice-worn country, varied with lakes, glaciers, snow peaks, and torrents. The glaciers had ploughed out huge furrows from Ata in the south to Shoshi Dzong in the north, a distance of fifty miles. Now all that remained of the ice-sheet were a few score disconnected glaciers, hidden away in deep ghylls, or stranded high up on the rock face; and the carved valleys bright with flowers.

But even that picture conjured up at the sight of these present lakes represented only the latest phase before the ice finally retreated.

At a still earlier period, at the height of the glacial epoch, the ice belt stretched from the Tsangpo to the Yangtse in a belt 250 miles long and 100 miles wide, completely cutting off interior Tibet from the warm southern plains. It was owing to the primary excavating work of the glaciers from this ice belt that the rivers of Tibet were later enabled to escape from the plateau between the eastern end of the Himalaya and the mountains of Szechuan.

On September 19th we returned to Lhagu, and next day reached Shugden Gompa.

CHAPTER XVII

LOST ON THE GREAT RANGE

'WELL, Tsumbi, you are not happy with me. I will pay you your wages to date and give you money for the road, and you can start for your home in Darjeeling to-morrow. You had better go by the big road to Lho Dzong, and thence by the Lhasa road and Gyantse; you will thus meet with travellers, and be quite safe.' It was the morning after our return to Shugden. After breakfast I had sent for Tsumbi, who now stood before me.

He drew a deep breath, looked at me, and then looked down at the floor. He was smartly dressed, his black eye had mended, he bore no signs of the late fracas.

'If the *sahib* commands.' He hesitated and then: 'Please forgive me, *sahib*. I wish to stay with you. I will never get drunk again.'

'Tsumbi, you said that before. I warned you of the consequences.'

'*Sahib!*'—Tsumbi seized his ears and pulled them—'you may cut off my ears if ever I drink alcohol again while I am with you. It is a promise.'

It occurred to me that during the fight in the dzong my opportunities for cutting off Tsumbi's ears had been few and uncertain; nor would the loss of them have prevented him from knifing me. However, I relented. Since we left Sangachu Dzong his behaviour had been exemplary. But then so it had been for a time in June, when I had told him that I was thinking of sending him back with my companions and taking Pinzo as sirdar. The threat then worked wonders; but its efficacy had gradually worn off as it ceased to have any meaning when he found himself at Shugden Gompa.

But the end of my stay here was approaching. Perhaps it would be better for us to stick together. I gave Tsumbi a final word of warning, and cancelled his dismissal. He was greatly relieved and thoroughly chastened. It is only fair to say that he kept his promise; for the next three months not a drop of alcohol passed his lips.

I spent another six days at Shugden Gompa. Late as it was—there were flowers in the marsh, notably a robust form of *Primula sikkimensis*, and a tall yellow flowered Cremanthodium, which colonized the almost stagnant streams.

Other alterations were taking place. All the babblers seemed to have departed, and the choughs went about in great flocks. More than once I flushed a solitary snipe.

The abbot called on me almost daily. Once he came in full canonicals, his red gown with the silver braid, and yellow cloth helmet on his head. I photographed him with an artificial scowl on his seraphic features. He brought me a sprig of dry Cypress, and a piece of scented wood, something like cigar-box wood, but probably also Cypress. He said that it came from Ka-Kar-Po, a sacred mountain on the Yunnan border, around which I had myself once made pilgrimage.[1] I had given the abbot many presents, and intended to give him more. But I was slightly shocked when he asked point blank for a present in return for allowing me to photograph him.

There was more activity in the monastery too. Both monks and laymen—now that the barley was reaped—were busy repairing roofs, whitewashing walls, and *chortens*.

And still Shugden Gompa continued to split the vapours of the air in twain, even as Moses divided the waters of the Red Sea. Clouds surged over the line of snow crests to the south, and the wind tearing off the fringes, sent them scurrying northwards, till there was a great blue trough of sky between the smoking glaciers and the distant violet ranges of the north. But there was a new menace in the wind. Occasionally I awoke to see the surrounding peaks powdered with fresh snow.

It is at this season that trade is brisk in Tibet, and the dzong, as a district headquarters, began to hum with activity. A large caravan carrying salt from Pashu to Sangachu Dzong arrived; and the drivers, both men and women, slept in the dzong, while the bags of salt were stored in the large ground-floor room of the main block. For fresh transport had to be provided here, and it took several days to collect. In this part of Tibet the goods exchanged are 'tea' and rice from Zayul, for salt and *tsamba* from Chamdo district. Two bags of rice will purchase a bag of salt. On my way to Sangachu Dzong I had passed a tea caravan from Rima, northward bound, comprising 600 yak and ponies. Now there came to Shugden Gompa, 100 yak laden with salt bound for Rima. Salt is another article which might well be sent up the Lohit valley from Sadiya. The nearest source of supply to Rima are the brine wells at Yakalo on the Mekong, about twenty-four days' journey over high mountain passes, with two great rivers to cross. The road is passable for pack animals, but the Lohit valley is infinitely easier, and it would not be difficult to restore the mule road built in 1913.

And now a new peril to my seeds made its appearance. Seed collecting sounds an easy occupation; anyone can collect seeds, you say. But apart from the physical difficulty of gathering seeds which ripen any time between August and December, over so wide an area, there are certain special difficulties. Slugs and larvae play havoc with them. Tragopans and other birds eat them. Rain washes them from their capsules, wind scatters them; and some are mechanically propelled from their capsules. Even those which survive all these perils may be buried under a blanket of snow, and lost beyond recall. But once I had spread the seeds out to dry on the floor of my own room I thought they were safe. Not a bit of it. A skirmish of mice descended upon them after dark, made hay with a bundle of Nomocharis capsules, and chewed up the seeds. I had to get some more and put them in a safer place. But that did not rid me of the plague, and mice continued to haunt my room, sometimes running over my face at night; I would awake with a start, thinking the devil had got me.

The day before we were to start on our last excursion, Tsumbi, Tashi, and Kele, dressed in their best, appeared before me and asked for a holiday. They wished to burn incense at the shrine of a famous image of Buddha. This shrine stood on the brink of a 200-foot cliff, almost opposite the dzong, and separated from it only by the gorge of the Tzengu Chu. Having performed this ceremony of burning incense, the men returned well pleased with themselves.

On September 27th I started once more up the Tzengu Chu. My plan was to cross the northern crest by another pass, and without descending to the Salween, reach Shoshi Dzong; my object, to get observations for latitude, between Shugden Gompa and the Salween, as near the passes as possible.

The mountain sides were brilliant with autumn colours, which showed up vividly against the sombre green of Juniper and Rhododendrons. The August flood had subsided, and the water was a clear ultramarine; Gentians spread a carpet of Cambridge blue over the bogs. I noticed an occasional white-flowered specimen. Arrived at the uppermost yak camp, we found the place deserted; but there was enough yak dung about to keep us in fuel. It was a dull cold evening, with a raw wind flicking us; no stars were visible. I had now to decide on a route. The four coolies with me, after consulting with Tsumbi, suggested that we make for Ru, and to this I consented. Now Ru lay somewhere to the west. Instead of crossing the Traki La, we turned aside up a broad valley which led straight to the foot of the glaciers, and camped on the highest alp; it was a short march, but it placed us in a capital position for crossing the range next day.

A stream of cold clear water from the glacier gurgled past our camping ground; behind us, and across the stream, bare screes buttressed the shattered ridges. An astonishing variety of flowers lined the bottom of the valley, including Saxifrages, Gentians, Aster, Dracocephalum, and Saussurea; they glistened like spar. Scattered tuffets of a dwarf bronze-leafed Rhododendron bulged up. It was cold and windy at sunset, but the sky was clear for half an hour, enabling me to get an observation for latitude.

The last day of September was full of interest to me. My guides started gaily up the stony valley, and we soon reached a kind of half-circus at its head. Vast mounds of gravel told of former glaciers, but only one glacier remained. We gazed up at the semi-circle of cliffs which enclosed us, seeking a pass; the men halted, irresolute. The cliffs were sheer; a snow cornice overhung the only visible col, at the head of a steep couloir. Formidable screes blocked every approach.

'Which way, Tsumbi?' I asked.

He shook his head: 'The guides don't know the way. They have never been here before.'

Then Tsumbi, Tashi and I started on a tour of discovery. Our sturdy ponies climbed up the great banks of gravel, and we went on to the extreme head of the valley till we could see every gap in the castellated wall; but there was no way out there. Only the couloir remained. Dismounting, we climbed a thousand feet, dragging our reluctant steeds. Slipping and slithering, puffing and blowing, foot by foot we scrambled up, till there was only the snow cornice between us and the top. This proved easy to break down, and we found ourselves on a broad saddle. The slope eased off, and we stood on a dazzling

(*Above*) GENTIANS AT SHUGDEN GOMPA
(*Below*) STELLERA CHAMAEJASME

snowfield, gazing down a valley. The sun shone on arcs of gleaming water; the valley was dotted with yak grazing.

'Hurrah! Ru must be down there. Go back, Tashi, and signal to the men to come up.' However, we both went back, sinking over our ankles in the soft snow, till we reached the head of the couloir. But the Tibetans were already on their way, the sure-footed yak plodding steadily up the slope, breathing heavily. It was as well. We had hardly started down the snowfield, when without warning the sky darkened, and a hurricane of driving snow almost blinded us, completely blotting out the scene. The slope steepened rapidly. The yak made light of it, but the ponies were floundering. The storm abated—it was like a frozen typhoon while it lasted—and now to our consternation we observed what the slope had hidden from us; the snowfield ended in a glacier. Presently we were on rotten ice. I kept a sharp look out for crevasses, but there appeared to be no danger. Suddenly one of the Tibetans pointed. Across the snowfield a number of antelope* were going up towards the pass in single file. I counted eleven, three of them quite young. At first they walked: but catching sight of us they began to leap and run, and soon disappeared behind a ridge of snow.

Having crossed the glacier we descended the moraines to a tiny lake at the foot of the ice wall, and were soon picking our way amongst the great boulders which strewed the valley. The grazing yak looked at us curiously, approached sniffing, and jerked aside; but unfortunately they could give us no hint as to where they came from, or where Ru was situated. Another valley joined ours, and someone said Ru lay up this valley, and over another pass. It was snowing again now, and we spent an uncomfortable hour searching for a non-existent yak camp, and a problematical village. Before we turned the last corner, I looked back up the valley we had first descended, and could hardly believe my eyes. The head of the valley was shut in by the usual high wall of rock, and no less than three glaciers converged on the lakelet at the foot of the wall, draping it with an icy curtain. A slight gap in the sierra was all that indicated a pass; it seemed incredible that with our laden yak and ponies we had actually crossed the range here. Seen from this side no one would ever have suggested the possibility of crossing. As a matter of fact, the pass, called the Ru La, is rarely used

---

* *Gowa picticaudata*, probably the Tibetan gazelle.

even by hunters; the herds do not use it at all. I estimated its height as about 18,500 feet.

We had now been climbing over rough country for six hours; and I was determined to camp at the first convenient pasture. We soon found such a place, and made ourselves as comfortable as the unpleasant weather would permit. I was cold and hungry, but very pleased at having got over this difficult pass. Also I had collected several interesting plants. On the high screes were clumps of the lovely pale lavender blue *Delphinium Brunonianum*. The plant itself is not six inches high, but the papery flowers are very large. Its interest lies in the fact that it is one of the highest flowering plants met with, ascending to over 17,000 feet, with species of woolly Saussurea and stiff clumps of Sedum. It never appears before August or flowers before September, and must spend at least seven months underground and four months under snow.

Hardly less interesting is *Primula macrophylla*, probably the highest Primula in the world; it does not occur much below 17,000 feet. It was very abundant in the soft friable earth and sand which forms the beds of glacier streams, and was now in ripe fruit. Still in bloom were an aquatic Cremanthodium, with kidney-shaped leaves, the shrill yellow *Draba alpina*, and Larkspur.

The grazing yak, astonished to see our green tents go up, and curious to examine them more closely, crowded round. One of them, more bloated than a yak should be, was seized by the Tibetans, and I watched with interest while a crude surgical operation, which involved an incision in the neck and a little blood-letting, was performed. It appeared that unless this was done, the yak would die, and its meat be unfit for consumption. The yak did not belong to us: but apparently it was the custom to perform this veterinary operation on your neighbour's kine.

A bitter white frost lay on the ground next morning, and the mountains loomed up dully through a heavy mist. We were still at 16,000 feet, and however physically tired I may be, my mind is always too active to permit of much sleep at these elevations. It is difficult to get mentally tired by yourself—or so I find it. We marched down the valley hoping to reach Ru, or at any rate a yak camp. Opposite a wide-mouthed valley to the south-east, we met two herdsmen. I had an odd feeling that I had been here before, as though in a former existence. But a little later we came to the place where we had camped on August 11th. The big side valley led to the Traki La. When we asked the herdsmen the way to Ru, they laughed and pointed up the valley we had just come down. We would have to cross a pass at its head, more to the west. The herdsmen took us to their camp a mile farther down the valley. There were seven black tents; all the nomads had gathered together preparatory to a descent into milder regions. Summer was over in the high valleys.

Entering the first tent, we were warmly greeted by our friends of the Traki La. We rested, and drank hot milk while the herdsmen prepared bundles of firewood for us. I also bought the empty carcass of a goat, for which I paid 15 *tankas*, equivalent to about 3s. 6d. All the best parts had gone. When we left, the herdsmen supplied us with four coolies, each of whom carried a bundle of firewood, enough for one night; for we had to camp at the foot of another high pass. The headman himself accompanied us as guide.

Ascending a valley immediately above the pasture, we entered the bleak region again, and camped in sight of a mountain which smouldered red in the setting sun.

CHAPTER XVIII

THE ROAD TO RU

ABOVE our camp the valley expanded on to a bare plateau at a height of 18,000 feet. It was as well we had a guide with us, for it looked as though we could cross the range anywhere within a space of two miles. We could have reached the rim of the amphitheatre anywhere; but only one valley led down to Shoshi Dzong. There were no outstanding peaks in any direction. Passing the red mountain, we reached one which was dazzling white, and crunched over a broad saddle strewn with crystals of calcite. Near the pass was a small glacier and a lake. A saw-edged wind rasped our faces, flaying the skin. Then began the descent of another interminable valley. All day we marched, descending gradually from a lifeless region of rock and snow to alpine pastures, from alpine pastures to scrub. However, the monotony was enlivened late in the afternoon by an incident. The path here skirted the hillside a hundred feet above the bottom of the valley, where a narrow strip of pasture showed. A quarter of a mile ahead stood three black tents; and I noted the tall lanky figure of a Kampa herdsman standing outside one of them. Two dogs now detached themselves from the encampment, and bounding up the slope were lost to view behind a shoulder. I was on foot, my party some distance behind. Herdsmen and dogs—usually tied up—were familiar objects, and I paid no particular attention to them until, suddenly rounding a corner, I saw the great angular head of a Tibetan mastiff poked up from behind a rock, not twenty yards away. It looked at me without a trace of friendliness in its bloodshot eyes, and I suddenly realized my danger. Being quite unarmed, I hastily picked up the largest stone I could find and retreated backwards, shouting to the herdsman below. Then the second dog popped up its head. They were huge, hungry-looking brutes, weighing perhaps 150 lb. each; these mastiffs make a blind rush, and in the first onslaught will knock a man down. At that critical moment there came a most welcome diversion. Some of the yak drivers farther back had seen and taken in the whole incident. Suddenly I heard a man running behind me, and a stone whistled over my head. 'Pönpo,' panted a stout fellow, 'look out! dogs!' He dashed past me, with drawn sword, and the mastiffs began to retreat. More men arrived and my servants, shouting angrily to the herdsman, rushed up. Reluctantly the dogs drew off, were captured by the surly Kampa—who throughout the proceedings had displayed a truly remarkable detachment—and tied up. I felt greatly relieved; I had had a good fright.

But the incident did not end there. Both my servants and the yak drivers were furious at this wanton attack on peaceful travellers. I was angry myself; we had not gone near the tents, and anyhow it was a churlish way of greeting strangers. They all ran down the hillside, and were on top of the Kampa before he knew what was happening. I had remained on the path, thinking that they were only going to abuse him, as possibly they were; but he showed fight, and the men, now thoroughly roused, set about him. An ugly situation now arose. Two other men came out of the tents and picked up stones; the yak drivers, going to the rescue of my men, were greeted by a fusillade, and seeing that a fracas was unavoidable, I also ran down the slope. A woman screamed and danced round the combatants: I seized a handy billet of wood, and threatened to beat either of the two men who were skirmishing on the flanks, should they join in the melee. I also hit one man to make him drop a stone he had picked up, keeping a wary eye open lest he should try to loose the dogs on us again. Otherwise I saw no good reason why the Kampa should not get the beating he had so richly deserved. The battle was now completely localized. The tall Kampa fought three of my men valiantly. His face was covered with blood from a cut over one eye, his hair, frequently seized by his assailants, hung matted and unkempt, his clothes were torn. The men pinioned his arms, pummelled him, and tried to throw him. He fought them silently, doggedly, and Kele at least carried marks of his displeasure. But three to one is long odds, and he had no chance; my men were too strong. He was seized, his arms bound with leather thongs borrowed from the yak drivers, and he was pushed and pulled up the hill on to the path, a prisoner. One of the young men, stung to action possibly by the imprecations of the virago, dashed up the hill to lead a rescue party; but Chimi drew his sword and turning on him suddenly, sent him headlong in flight. It was the first really comic interlude, and I could not help laughing. The valiant Chimi—what a hero! The youth who had slunk away blubbering when Tsumbi got drunk and abused him! Needless to say it was Tsumbi, Kele, and a yak driver who had captured the prisoner. The young man, deeply concerned about his father's safety, now dived into one of the tents.

'They have gone to get their guns,' remarked the fearful Chimi.

'More likely to loose the dogs,' said I.

But the man emerged from the tent, not with gun or dog, but with a tray of *tsamba* and a large round of butter. Shouting to us, he held up his peace offering. But we would have none of it, and marched off our prisoner, who now went quietly; indeed he could do nothing else. We did however allow one of his relations inside the cordon; it was his old mother who, hearing that he would not return that night (we were adamant on the point), brought him a cloak and some spirit. Possibly his family thought they were seeing the last of him for many years; some of them certainly behaved like it, for a great wailing arose from the stricken camp. But the prisoner himself was philosophical, and only asked to be allowed to wipe the blood from his face. He was a saturnine man, not ill-looking, with a hawk-like nose and a cast in his eye.

We now resumed our march down the valley, but were told we could not reach Shoshi Dzong that night. Tempers cooled rapidly—indeed it was a chilly evening. The prisoner laughed and chatted with his captors. I took Tsumbi aside; he was still indignant as befitted a sirdar.

'We'll let the man go to-morrow morning,' I said.

'No, *sahib*. He behaved very badly. We will hand him over to the magistrate at Shoshi Dzong and have him punished.'

'I don't want to make trouble, Tsumbi, it isn't worth it; and there is no magistrate at Shoshi Dzong. Besides, you gave him a good beating, you know!'

Tsumbi beamed.

At dusk we camped. The prisoner, now unbound, helped to collect firewood. He spent a cheerful evening with his former enemies, and early next morning Tsumbi pleaded for him! I willingly let him go and he ransomed himself with a chunk of butter, apologizing for having loosed the dogs on us. His excuse was that travellers never used that pass, and he thought we must be robbers!

The valley grew still narrower, till we were wading in the torrent which at one place filled its bed from cliff to cliff. At noon we emerged from between grim portals into the main valley, a couple of miles below Shoshi Dzong. On the way north down the valley I had scarcely noticed this crack in the mountains.

Crossing the river by a wooden bridge, we reached the dzong, now more deserted than ever. Even the caretaker had gone; and we had some difficulty in getting any firewood.

On October 4th we headed south for the Poyü La, meeting vast numbers of yak, sheep and goats which, escorted by shaggy mastiffs (some of them led on chains), and herdsmen with their worldly goods, were moving down from the high pastures. As we turned aside to reach the pass, one of the men pointed to a path which continued up the main valley.

'Where does that go?' I asked.

'The road to Ru,' he replied.

After that, to suggest that is Mrs. Harris would be merely frivolous.

We camped by the Poyü Tso. The last of the nomads had not yet departed; there were encampments on both sides of the lake.

It was a brilliant night, with an amber moon rising over the jagged ranges of Pomé: the first really clear night since we left Shugden Gompa. But this did not last. We awoke next morning in a thick mist. There was fresh snow all round us, and the wind blew fiercely from the south, bringing first snow and then rain. It was a cold and joyless ride back to Shugden Gompa; and the storm lasted another day.

I had been absent only nine days, but it had brought the time of departure very much nearer. There was no object in staying on here after the middle of October, by which time I should have collected all the seeds I wanted; and I debated whether to return by Sangachu Dzong and the eastern Zayul river or whether to try and force my way over the Ata Kang La and return through the Mishmi Hills. By following the former route I would be able to botanize on the unexplored mountains east of the Lohit river; by following the latter to explore the Rong Tö Chu to its source and botanize at the top end of the Mishmi Hills. Both programmes were attractive, and I did not finally make up my mind until one evening the fat headman of Modung, bulking larger than ever in sheepskin coat and fur cap, waddled in. He had come to Shugden Gompa on business, and intended to return to Modung within a week. I decided to accompany him.

But before I could leave Shugden Gompa I must collect seeds of all the beautiful alpine flowers I had found, especially the Gentians, the Dragonheads (Dracocephalum), mahogany flowered Trollius, and certain Rhododendrons. I sent Kele and Chimi to the Zo La, half way to Sangachu Dzong, on a special mission to collect the Trollius and a dwarf Rhododendron; while Tashi and I climbed Ningri Tangor on a biting day. When all the seeds were gathered in and spread out to dry in my room I started to pack. As I emptied my boxes I felt as excited as a schoolboy at the end of term emptying his locker. For months I had been living alone in a world apart from the one to which both birth, upbringing, and tradition had assigned me; a world of my own choosing, yet none the less alien. No sign from that other familiar world had reached me for five months; I knew nothing of what had taken place in it since February, nearly eight months back. It might have disappeared in chaos for aught I knew; its utter annihilation could have no immediate repercussion here. And now I was going home, back to that familiar world. No matter that within six months I should feel as much an intruder there as I now began to feel here. I was stale. I wanted to rest, to see people, and cities, to talk; more than anything I wanted to get down to lower levels, to sleep. At the thought of going home all the tension of the past seven months seemed to relax; I felt almost light-hearted. Nevertheless, I had no illusions concerning the return journey. The hardest part was yet to come—the crossing of the great range, plant-hunting on the snow-bound hills, where no flowers were to be seen, winter. I did not expect to reach India till after Christmas, which was still ten weeks off. While packing went on several small children hung about under my window picking up the junk I occasionally cast out. Callers also, both lay and ecclesiastic, came with an eye to the discard. I fixed October 18th as the day of departure, though Kele and Chimi had not yet returned from the Zo La. Three days before we were due to start it began to snow and we awoke to see the whole countryside white. Although much of the snow melted during the day, more fell that night, and the mountaintops put on their winter caps. Kyipu shook his head; if this continued, he said, we could not cross the Ata Kang La. The abbot announced that special prayers for fine weather would be offered up in the temple. Taking my usual walk that evening, I saw no less than eight hares; but it was cold comfort, for the local soothsayer could pronounce no omen, good or bad, on the strength of it. On the 17th the two men returned from the Zo La. They had intended to return the previous day but, owing to a violent snowstorm, had been afraid to leave their tent. They reported deep snow on the pass: but they had got the seeds I wanted.

CHAPTER XIX

FAREWELL TO THE MONASTERY

OCTOBER 18TH was a red letter day: it was like breaking up for holidays. To start south again was to make steps for India and home. I was glad to be leaving this bare grassland for the forests, and the warmth of India.

I rose early to pack my last boxes of specimens and make all final preparations. I was in a hurry to start. For the night had been warm, a few degrees above freezing, and I was afraid of snow; a heavy fall now would block the pass of Ata Kang La and cut me off from Zayul.

Coolies began to arrive from villages round the lake, from Yangoong and Changra, and even from Arig, which is on the far side. These crossed the lake at its narrow neck, riding on ponies; for the lake shrinks in autumn and grows shallow, while in winter it freezes solid, so that men and ponies can cross it anywhere. I was told that in summer they used dugout canoes for transit; but I saw none, only rafts, on the western arm.

The coolies took two hours to divide up loads and tie them on the yaks. Meanwhile I said good-bye to the monks. The abbot came early with two wooden bowls as gifts. They were varnished and their bases painted with gold. In return, I presented him with my folding chair. This pleased him so much that he went away and returned with a third bowl of a slightly different shape. He gave me that also, explaining that ordinary people were not allowed to drink out of cups that shape; it was a privilege, reserved to abbots, nobles, and me. What interested me was that in the bottom of each bowl a small pat of butter was stuck. I could not find why this was done. It may have been to ward off poison, or as a symbol of plenty.

Kele, who having learned that his young daughter was ill, was returning to his shop in Chamdo, also brought me two wooden cups. He and the abbot each presented me with a ritual scarf and asked me to wear this on the road. These scarfs were of very coarse stuff; but they averted evil, I was told, and ensured courteous treatment from those I met. Kele was drunk from the night before. He was voluble with the sadness of parting. I gave him a stores box and a 'Beatrice' stove, for which he said he could get oil in Chamdo, a statement which surprised me. I also gave him six letters to send to England via Lhasa. They reached their destination, despite the distance and the state he was in.

The Dzongpön, as ever practical, brought food for the road—rice and barley flour, in return for which I gave him several tins, some of which were not empty. For the monk who had showed me the rose Primula I reserved an empty stores box of three-ply wood with padlock.

Meanwhile prayers were being offered in the monastery for our safety, and smoke rose from a dozen pyres, which the monks had lighted. The abbot asked me when I would come back. Tsumbi and Tashi were invested with scarves, and they gave scarves to Kele, the abbot and the Dzongpön. Then the abbot blessed us, and we could have started at once, if the coolies had been ready.

We set off at 10.30, straggling into line. I rode down the steep path to the lakeside for the last time. Half a dozen old women and young children stood outside their huts beside the dzong and waved goodbye. They smiled and called to me to come back. At the foot of the hill we all drank a stirrup cup with the Dzongpön, and his aged wife pressed a hemispherical brick of Chinese brown sugar into Tsumbi's hand. Then at last we got away.

The mist which had hung all the morning over the lake rose in the sun's heat, leaving a white blade of cloud which severed the peak of Dorje Tsengen from its base, so that it seemed afloat. The sky, cold and turquoise, was reflected by the lake; while the valley was yellow, from the dusty cinnamon of rocks to the honey colour of stubble fields.

I was in high spirits. Our chances of reaching the Rong Tö valley were good, though I had left it late to cross the Ata Kang La. They told me it was often closed by the first week in October; soft snow hiding the crevasses on the glacier. I required the seed of three plants on this side of the pass. If I got them I considered my time at Shugden Gompa would have been well spent; but if I failed, I might regret it always.

We halted at Yatsa for a late lunch, the yak travelling leisurely. The coolies said we could not reach Lhagu that day and wanted to stop the night; but as the moon was full, I answered that we would go on, even if we did not get in before dark. That made them hurry and we arrived before dark, about 5 p.m.

FAREWELL TO THE MONASTERY

DORJE TSENGEN

On our way I had one success which I took as an omen. There is a sheltered cliff above the path in one place, clothed with thick bushes and even small trees of *Picea lichiangensis*. Among these bushes are many Rhododendrons, chiefly *R. sigillatum,* but also a species which from its leaves I judged to be close kin to *R. vellereum** though I had not seen the flowers.

This Rhododendron had exhausted itself in 1932 in a glut of bloom, so that the following, i.e. 1933, it had no strength to flower at all. Its branches were covered with the dead brown trusses of the year before, but though I had searched these bushes several times, I had not found one green capsule. As we climbed the rough path, I decided I would have a last try, and slipping from my pony I began the search. There were many bushes above and below the path; but though I examined all, I found nothing. As hope ebbed, I looked more perfunctorily. I grew certain that no bush had flowered this year and I was wasting my time, that I must acknowledge failure. Then I saw a bush above me with green ripening capsules and involuntarily gave a whoop of joy. I climbed to it quickly and pulled off every fruit. The capsules were riddled with caterpillars, which had eaten most of the seed. But I knew there were some left and hoped it would be enough, for Rhododendron seed is very small. I searched for more; but though I examined many more bushes, I did not find a single other capsule. Only one bush, apparently, from two or three hundred had flowered this year. But I had pitched on it and was well satisfied. From the worm-eaten capsules I extracted, by drying, a pinch of fertile seed, which germinated readily in England. This was the first of my three seeds wanted. The other two were *Primula szechuanica* and a dwarf Monkshood.

Lhagu was cold; an icy wind blowing from the glaciers. The people agreed that we could still cross the Ata Kang La, but there was no time to spare. Their local tradition says, while there's no snow in Lhagu the pass is open. There was no snow, and the sprinkling that fell during the night soon melted.

We could not cross the pass that day; we had to stake on the weather holding, camp the night at the glacier foot and rush both passes the third day. Up the valley above Lhagu we found the headman of Modung also waiting to cross the pass, and camped near him.

---

*   *R. vellereum* is a beautiful species of the 'Taliense' series which I discovered in the gorge of Tsangpo river in 1924. It is now in cultivation.

FAREWELL TO THE MONASTERY

It was two when we camped, and I spent the remaining hours of daylight collecting seed of various alpines, especially *Primula szechuanica*, the Monkshood and Gentians. I had got what I wanted.

The morning had been sunny; but clouds came up in the afternoon and hung in a low fringe along the snow range. A stiff breeze, shredding these clouds, brought them down the glaciers in a fine drizzle. Though I could almost see the top of the pass, I knew that if it snowed during the night we could never cross those few miles of ice and find our way down the *séracs* on the other side.

I lay on the hard turf picking off the capsules of the Lhagu Gentian until it was too dark for me to see. Then I went inside my tent, which seemed colder than outside. All yak and riding ponies had been taken back to Lhagu. Coolies and guides were gathered in camp around fires they had built. They piled my boxes round them as shelter against the bitter wind.

The stars were brilliant in the early night; but a mist creeping up soon blotted them from sight. We all went to bed at 9.30.

I awoke early with the uneasy feeling that the air was too warm. The thermometer showed that during the night the temperature had not fallen below freezing point. Pale mist filled the valley, hiding alike the mountains and the warmer colours of earth. Wind whirled a fine snow from the glacier and the camp was quickly covered with a white film. When I drank my cup of tea at 5.30 I saw several larks hopping in the snow.

The day did not look promising; but the coolies set about preparations with such good grace that I took heart. By 7.30 we began climbing the steep snout of the glacier. The ice was covered with nearly a foot of soft snow into which we sank each step we took; and the wind, coming head on, blew frozen snowdust steadily in our faces. The sun's disk shone palely through the scud. Every moment I expected it to break through; but it never did.

Ascending the glacier was hard work, but we reached the top in three hours. Snow gave way to drizzle and the temperature rose. Through the mist we looked down at the Ata Kang La south glacier. Three months before when we had climbed it, it had been a long smooth slope of firm snow. Now it was cracked and scarred into a multitude of sharp *séracs* and deep crevasses filled partly with snow.

The route we had followed in July was impassable. We had to make a precarious way down the right bank.

I had with me two one-hundred foot ropes. The two leading parties would be roped, and the remaining thirty-five coolies follow unroped. We saw a track in the snow, marked by coolies coming up from Ata and I reckoned that if they could cross the glacier safely, we also could. All the same, I was glad of the two ropes. Tashi, a coolie and myself were on the front rope, the coolie bringing up the rear; while Tsumbi led the second string with two coolies behind him. Nothing serious happened. I put my foot in a crevasse once and fell. Others had falls. But no one was injured. On the way down we passed six coolies coming from Ata, carrying wooden churns, gun stocks and other forest products.

After descending some hundreds of feet, we left the glacier where it was badly fractured and climbed the cliffs on the right. The path then plunged steeply down to the edge of the glacier again. This was worse than anything before. We had to cross the glacier—which was about a thousand yards in breadth at this point—to reach the moraine on the left bank near our old Glacier Camp. The coolies who had just crossed had trodden a path in the snow, which we followed. The snow had not yet fallen so deeply that we could not see the crevasses. There was no danger of falling into them unless we slipped on the ice. I was the only one with nails in my boots. But I reckoned that what they lacked in equipment, the coolies made up in skill. At any rate we got across without mishap.

There was no use in halting at Glacier Camp. Our clothes were wet and we had no firewood. We had to go on, cross the Cheti La and make Chutong Camp. The ascent to the Cheti La had always been unpleasant. Now, wet and exhausted, I hated it more than ever.

It was an enormous relief to stand upon the top at last, and look far down the Ata valley, banded with dark forest on either side with the soft clouds curdling overhead. No snow lay on the pass, nor down the slope below; and whereas, in the Lhagu valley the gentians had been seeding, here many were flowering. They formed great patches of prussian blue on the crisp turf. Growing beneath the Rhododendrons was another species with leafy heads of watery-blue flowers. Only one gentian (*G. Wardii*) or Blue Bubble, was in ripe fruit; I began to collect seed of it.

The descent to Chutong was easy; but, when we arrived, we found there was no water, nor even snow, as there had been in June. Men had to climb up towards the pass again, a thousand feet, to fetch a minimum of water.

While the coolies were preparing the camp, I went off to collect seeds; a penance after nine hours' climbing on rocks and glaciers and crossing two passes, but a necessity. In the course of an hour I made two important discoveries. The first, that a small leafed scarlet berried Lonicera I had found at Shugden Gompa was *not* the same as the Chutong small leafed, scarlet berried Lonicera; their leaves in reality being quite different, the one glossy, the other felted. The second, that the yellow-flowered Lonicera of Chutong had bluey-violet instead of orange berries and was therefore *not* the same as the orange berried Lonicera (*L. hispida*) of Shugden Gompa though otherwise closely resembling it.

I gave all the coolies a tot of rum, and we built a great fire, round which we had supper. We all went to bed early. The valley was warm and fragrant; the air, compared with the grit-laden air of the plateau, moist and clean.

We awoke in a dense mist, as though it were June again. Ata was invisible. But later the sun came out between showers. Looking towards the snow, I could see vivid patches of Berberis, in autumn scarlet. But the forest below was dark green, except for the red Vines which wrap round the tree trunks; and there was little autumn colour until we reached the open valley, where the leaves of Rhus and tree-of-heaven (Ailanthus) varied in every shade of red from scarlet to vermilion, brightening to champagne yellow and even unchanging green.

Half way to Ata at our old camping place we met the headman of Modung with his men, sitting round a fire eating. Like us they had had a hard time the previous day and were resting. We stopped and ate with them.

It was warm now; and from the trees came the noise of cicadas, like the grinding of scissors. What pleased me most was presently to hear the patter of raindrops on leaves, a sound I had not heard for months. For though there is rain on the plateau, there is no forest, and the only noise it makes is a sort of hissing on the hard earth. I thought with pleasure of the warm wet jungles of Assam, which now were no great distance off.

We sat down beneath the towering trees—there was a coarse-leafed yellow-flowered Salvia growing here forming a dense undergrowth—and one man drew a small knife from a sheath, dangling from a chatelaine at his waist, and cut slices from a chunk of pork. Another opened a bamboo box in which was butter wrapped in a leaf, and red pepper. They threw the meat into the fire, and picking it out in a few minutes, crisped on the outside, ate it almost raw. I did the same. It was the first fresh meat I had tasted for many days.

We continued the march to Ata together and reached it in the late afternoon. I slept in my old room and next morning walked over to see the headman, a rather greasy looking old man with tousled hair and sagging knees.

When I first entered the kitchen, I could see nothing for darkness, then I saw a pile of rags on the floor. The rags moved, got up and came over to me. It was a terrible, semi-human figure. I think it was a woman. I could scarcely see her face. It did not reflect the light. It was dark and unclean. She grinned at me, and muttered something. Then she went and lay down on the rags again.

The air was foetid and sour. I sat down by the fire as I used to do and one of the men came in and brought me a bowl of milk.

'Where's Tsering?' I asked.

'Dead, Pönpo!' the man said.

'Where's that pretty little girl who used to carry firewood?'

'She's dead, too.'

I thought of the woman lying on the floor, the bundle of stinking rags. Tsumbi came in at that moment to say the coolies were ready. It had been difficult to find coolies, but everything had now been settled.

'Why aren't there any coolies?'

'There's sickness in the village,' he said. 'Several have died.' The whole village had been stricken by an epidemic; the house where I had slept, and Tsering's house along with the rest.

I said good-bye quickly and went out into the open air. The last coolies were leaving with their loads. A boy was carrying my bedding. I looked at his face. It ran with pustular sores.

'Give that boy a box or something, if he *must* carry a load—but for God's sake don't give him my bedding.' I was feeling peevish again.

Tsumbi made the change, but without conviction. He seemed to think I was fussy. It's no good worrying about having slept in an infected house; but where you can prevent a boy with smallpox carrying your bedding, I think it only wise to do so.

It was a typical autumn day: all peaks hidden, thin clouds blowing up the sodden valley, and a constant drizzle falling. We rode slowly through the wet woods where dead bracken stood three feet high. Autumn here lasted longer than on the plateau: it was more like England. Even on this dismal day, the bright colours of the shrubs cheered me; I particularly noticed the shrill yellows of Litsaea. The corn was reaped, but there were standing fields of winter buckwheat.

We reached Modung early in the afternoon and the headman gave me a comfortable room, all ready prepared. He had put some Dahlias and Hollyhocks in a bottle as decoration, and in the evening he ordered a pig to be killed in my honour. Pigs are killed by strangling them with a rope and then cutting their throats.

I had to stop now for a few days to dry and pack the seeds I had collected since leaving Shugden Gompa. I spread them out on sheets of paper all over the room; about forty species.

I now tried to persuade the headman to accompany me on a journey to the source of the Rong Tö river and over the Kangri Karpo La into Pemako. He replied that as a young man he had once been to Sadiya and as far as Calcutta but now that he was getting old—he was 37— he had to be more careful of himself! Nevertheless he would consult the monks, who could pronounce by divination whether or not it was advisable to risk it. Anyway, would I wait for a few days, as he would have to collect coolies for me and have a new rope bridge twisted.

Next day a deputation of monks waited on me to ask for presents. These I gave them, expecting them to give a favourable reply to the headman. I also presented the headman with a bottle of rum, a box of candles and a cake of soap. He in turn brought me a blanket, a ham of pork, some rice, flour, and eggs.

I spent six days in Modung. The weather grew fine and the sun shone warm in a clear sky. Now that we had got down to lower levels, I slept much better and enjoyed life altogether.

One slight shock disturbed me. Two thousand feet above Modung the dwarf Iris grows freely. I had noticed its abundant flower in June. But when I climbed up, hoping to get seed, I found none. On my way down, however, I found a few plants of *Lilium Wardii* which were in fruit. Most were about two feet high and ended in a solitary capsule; but a few had two capsules, and one had three.

As the fine days slipped by, I grew impatient to start. But the headman kept putting me off, saying that he had not got the coolies.

Meanwhile my landlord from Ata came down to see me. He whispered something to Tsumbi, who looked shamefacedly at me.

'What is it, Tsumbi? What's he want?'

'His wife has six children, *sahib*. Every time he sleeps with her, she has another. He does not want any more. Can you give her some medicine which will prevent her having any more?

It surprised me to find that these people do not know of any forest berry, root, or leaf which will bring about abortion. I should have expected them to have some knowledge of contraception or abortion, if they had not clearly informed me that they knew nothing about it. Most civilized races do.

The new rope bridge was finished at last and coolies from Modung, Ata, and Suku warned for duty. I waited only for the headman to say whether he was coming or not. The monks had given the oracular reply that if he went with me to Sadiya, he would never return. This might mean that he would remain in India—dead or alive; or it might mean that he would die on the way back. He decided that he would not risk the whole journey, but would accompany me to Putsang, the outlaw village up the Rong Tö river, not a tenth of the distance, where he would hand me over to the headman. He might as well not have come at all.

I had to be content with that and the start was fixed for October 29th.

CHAPTER XX

THE PILGRIM'S PROGRESS

It was a gorgeous day. Looking due south down the steep rock-lined ravine, I saw a high snow-capped peak framed between cliffs; it was one of the sentinel peaks of the Mishmi Hills. There was the usual inertia of starting from rest; we got off late. The coolies were in two detachments. The main body, bound for Rongyul, carried my collections and excess baggage; a picked detachment accompanied me to Putsang. As we descended the gorge it grew noticeably warmer. We crossed the Ata river, now forty yards wide, at the mouth of the gorge, by the new rope bridge; the footbridge had been carried away. That evening we camped on the bank of the Rong Tö river, under a cliff on which were many flowering shrubs. There was a charming Clematis (*C. acuminata*) with cream-coloured thimble-shaped flowers; and both *Rhododendron manipurense* and *R. bullatum,* deceived by the fine weather, were opening their buds for the second time this year. Sprays of Cymbidium trailed over the cliff edge.

Looking down the sunlit valley of the Rong Tö next morning, I felt a momentary pang as I turned my back on it and faced the high peaks once more. But by the time we reached the mysterious village of Putsang, I had forgotten my homesickness. Kyipu commended me to the local people and took himself off with his following. But before he left he ascertained that it would be impossible for us to cross the Kangri Karpo La, at the head of the Rong Tö valley; we were too late by six weeks or more.

The village of Putsang comprises a score of châlets dotted about in a bay of the mountains, backed by forest. The mountains were almost as upright as the Fir trees, which seemed to grow one on top of the other in stripes; their bare sugar-loaf heads thrust themselves aggressively into the china blue sky.

The headman, a gentle young man, tall and slim, with pleasant features, told me the story of Putsang. During the wars with China, a quarter of a century ago, the independent Tibetan tribes of Szechuan were harried by the Chinese troops. Many left their houses and sought sanctuary farther away from the disputed territory. One party migrated into Zayul, where there was much land and little population. Beyond the farthest outpost of cultivation in the Rong Tö valley, they came upon the site of an abandoned village, and here they halted, calling the new settlement Putsang. Long before this time the original settlers, despairing of wringing any return from the hungry soil, had migrated up the valley, crossed the Kangri Karpo La and reached the Chimdru river, in the valley of the Dihang. But the path they had made through the jungle was now buried, and the newcomers made no attempt to follow them. They live as much by hunting as by agriculture; takin, musk deer, serow, and gooral all occur on the surrounding mountains. They pay no taxes to Rima, maintaining, quite unreasonably, that they ought not to be taxed. At present the tax-gatherers leave them alone.

The head lama, a venerable toothless old man of 70, welcomed me. He had come to Putsang from Draya, when the Chinese occupied Rima. He showed me the monastery, half a mile from the village; a peaceful place. The walls are of stone, but all the upper part is timber, the doorway wonderfully carved. He bewailed the fact that the monastery is not yet finished—the walls unpainted, and the temple empty. In front of the building was a lawn, so green and crisply trimmed that it might have been lately rolled and cut with a lawn mower. It was scrupulously tidy, though the wooden stalls round it, which housed thirty monks, suggested backyards in Hackney. Immediately beyond the monastery a large glacier torrent from the north entered the main river.

I was pleased with my reception at Putsang, having heard that these people were stand-offish and superior. That reputation is undeserved. The headman and the lama between them brought me four dozen eggs and some turnips, and promised transport for the next day. My coolies were a queer collection. There was one man with a hare lip, who spoke to me in Chinese, which he had learnt as a trader in Szechuan; his mother was Chinese. Another was undoubtedly the missing link. His hairy chest was immense, and he had long ape-like arms; his face, covered with hair, with high cheekbones and projecting muzzle, resembled a gorilla's, and indeed he made gorilla noises. He was a good-natured monster. There was also a hunter, a tall lithe man who strode along like a Colossus, bare kneed, with a huge basket on his back, much heavier than any of my boxes.

A KAMPA GIRL

For five days we toiled up the valley, the gradient growing ever steeper, the mountains higher; more sheer they could not become. From one spot a little above the river I could see the top of a peak on the Dri river divide, 19,000 feet high. That is to say it was almost sheer for over 10,000 feet; two miles. Every stream from that side, rising only a few miles south, finally leapt hundreds of feet to the main river; for the crest line of the Mishmi Hills almost overhangs the Rong Tö gorge. No wonder the Bebejiya Mishmis do not cross the mountains above Putsang, nor the Tibetan hunters cross into the headwaters of the Dibang. Between Putsang and our last camp up the river we crossed four big glacier torrents from the north; but we rarely caught a glimpse of a glacier. These torrents poured out of deep wounds in the mountainside, and it was impossible to see what lay behind. One knew that they were glacier torrents, none but glacier torrents—swift water charged with sharp grit, could have cleft these wedge-shaped slits. As the Rong Tö, here called the Zayul Ngu Chu, or for no particular reason, grew smaller, the rapids grew larger. The river, now blustering where formerly it was powerful, broke into a succession of cataracts interrupted by quiet pools of a cathedral luminous blueness.

I marked our progress towards the crest of the snow range by means of the sequence of Rhododendrons I saw in fruit. Below Putsang there were *R. virgatum, R. bullatum,* and *R. manipurense.* Next to appear were the bristly *R. vesiculiferum,* with pinkish-purple flowers blotched at the base and *R. fulvoides.* Still higher came the crimson flowered *R. tanastylum.* But much the most interesting Rhododendron was a solitary tree crowded with fruit which I found overhanging the river at our highest camp. It had all the appearance of being a natural hybrid, possibly between *R. tanastylum* and *R. vesiculiferum.* Above Putsang the gorge of the river was thickly forested on both sides, with Coniferous and broad leafed trees. Spruce replaced Pine; Oaks, Maples, species of Ilex and Spindle tree (*Euonymus*) abounded. Not less interesting was the undergrowth. Where the ground was soft and wet, a dense spreading growth of Bamboo made progress difficult; but elsewhere the undergrowth was largely herbaceous. Species of Strobilanthes were in bloom. A scarlet-clubbed Arum (Arisaema))—the large leaf tri-lobed, each lobe wedge-shaped—grew scattered with *Lilium giganteum.* A stiff heathery growth of *Vaccinium glauco-album* encrusted the sun-soaked granite cliffs. Its rigid leaves are polished sea green above, and white beneath, and it bears bunches of glaucous blue berries.

As we approached the high peaks, the valley showed obvious signs of becoming drier; I could tell this from the appearance of the forest, and from its composition. There were fewer epiphytes, but a fair number of woody climbing plants; epiphytes—that is to say plants which perch themselves up in the trees, quickly disappear as the atmosphere becomes drier. On the third night we camped on a rock platform overhanging a furious glacier torrent. It was a gloomy spot. The cliffs towered up on either side, leaving only a ribbon of sky visible. The trees almost met overhead and the dreadful roar of the torrent crashing into the abyss below banished sleep. Next morning our hunters left us to climb straight up the mountain. They carried a large supply of food, axes, rope, and ammunition, besides their long muzzle-loading cheek guns. Three lean yellow hunting dogs accompanied them. We continued up the valley, now climbing steeply, then right down several hundred feet into the brilliantly illuminated river bed, clambering over the boulders, and back into the dim forest, pushing between the bamboos. The coolies, with heads bent, plod steadily on in single file, smoking philosophically, sometimes exchanging remarks. There are many obstacles—fallen trees and such-like. Sometimes we can pass by stooping under them; sometimes with much grunting the coolies clamber over them; sometimes we cut a way round. The faint trail, marked by pilgrims who make the long journey round the snow range annually, is often scarcely visible in the rank growth. The lush knee-high vegetation wets us to the skin. Ticks crawl on to us, to bury a sabre-like instrument deep into soft flesh. These ticks are very troublesome; once they bite, whether we remove them or leave them, they make a bad sore. The more we pull them, the tighter they hold and the deeper in goes the instrument. The result is always an itching sore which dirt makes into a festering wound.

As we marched I checked off the miles as well as I could estimate them: for at every mile we plunged deeper into the unknown. Our average rate of progress up the valley was barely two miles an hour.

On the fifth day the gorge began to open out; the river divided into two equal branches, and we camped at the confluence. We had reached the zone of Larches, the highest of all trees. The headman now begged me to return. 'We have no more *tsamba*, Pönpo. We cannot reach the pass; the snow is very deep.' I said we would remain here one day, and then start back. Had there been no snow it would have taken us another three days to reach the top of the Kangri Karpo La. From there to Shingki Dzong, the first inhabited spot on the Chimdru river, was seven days' march—making altogether fifteen days from Putsang.[1] We cut down the undergrowth and made a clearing in the forest for our camp. Through the naked trees we could see the harsh outline of high peaks, dark against a gunmetal sky faintly luminous with stars.

The following day I explored a short distance up the north branch of the river. About half a mile above our camp the torrent flung itself over a smooth granite cliff into a narrow chasm a hundred feet deep; one could almost stride across it. Above the cliff the valley broadened, the torrent gurgled placidly over a gravel bed, and we waded over knee-deep. Bare granite cliffs bounded the valley on either side. In the direction of Pomé I saw a snow-clad mountain. We put up some duck here. When we got back to camp we heard that the hunters had shot three takin. We started back next morning, but only marched four miles. Here the path traversed an open hillside, and I wanted to camp, in order to take a latitude observation. So narrow was the valley, however, that it was practically impossible to see all the necessary stars for a satisfactory observation. It was a brilliant night. As soon as the shadow of the western range fell across the valley, the ground glittered with frost, and the stars, pale in the momentary sunset glow, suddenly crystallized out of the rapidly chilling sky like sharp minerals. The weird shapes of mountains menaced us, cutting us off from a world of familiar sounds; the silence was only broken by the snapping of a rock or the hoot of an owl. With frozen fingers and numb feet, I took the required observations, then crept shivering into my tent.

We accomplished the few remaining miles to our number four camp with half the coolies, in two relays, next morning; the other half having formed a working party to bring in the dead takin. Having the afternoon free, I climbed a high cliff above the rushing torrent, and saw away to the north-east a snow peak on the great range.

That evening the hunters returned, and there was jubilation in camp. They had shot a large bull takin, a female, and a calf. It is becoming difficult to get these animals. They are hunted by the tribes—Abors, Mishmis and Bebejiya from the Indian side of the frontier, and by the Tibetans in Zayul and Pemako. There is no close season. Females and young are slaughtered indiscriminately. Only the fact that the takin country is so precipitous, so covered with dense forests, and so sparsely populated saves them from extermination. Takin, though not common, are abundant locally. They go about in herds, being usually found above the forest line in summer, but descending to lower levels in the winter. This particular species at least—*Budorcas taxicolor*—inhabits forested regions rich in Bamboos, and is confined to a belt between 8000 and 14,000 feet, descending into the river gorges.

The hunters had ripped the skins from the carcases and cut off the head and limbs; they and the coolies brought down all the meat, five basket loads. Wooden tables—one might say altars for burnt sacrifices—were quickly built, the meat piled on and hung round them in a curtain of strips, and smoking fires lit. After dark, the leaping flames lit up a strange scene. A circle of shaggy-haired men sat on the ground, under the fan vaulting of the great Spruces, which formed our leafy roof, and gnawed bones; the light flickered on the gnarled and ruddy trunks of the tree Rhododendrons. Hungry dogs still prowled round, licking bloody chops after scavenging in the offal. Through the bushes the river, flecked with red, was seen to disappear into a darkness as mysterious and final as that which swallowed up the Styx. We now had much more to carry home than we had started out with. The skins, skeletons and meat made three immensely heavy loads. Some of the coolies were carrying not less than 120 lb. apiece. I thought we should go slowly; but the headman said that on the contrary they would travel fast; for though they had meat, they had finished their *tsamba*, which is to a Tibetan what chupatties are to an Indian and bread to an Englishman.

That day the weather broke. Grey skies, a sulky drizzle and a thin deposit of snow on the Fir trees. All day we slogged through the dripping forest, a savage wind tearing the big Maple leaves from their last hold and slapping them like sheets of wet brown paper in our faces. On November 10th, after a long wet march, we camped at dusk in the river bed. It was cold and clammy. The Zayal Ngu Chu, now something more than a torrent again, and clouded with mud, filled the air with spray. After a brief night's rest we were on the trail again. In the night the snow crept down to 9000 feet. A party of hunters whom we had passed on their way up the day before, joined us, saying that the snow was too deep up above; they also were returning to the village. At last, cold, tired and dispirited, we suddenly emerged from the sodden forest into the clearing where stood the tranquil monastery of Draya. In this peaceful haven, with the mountains all round us shrouded in cloud, we sat down to rest for ten minutes; then went on to the village of Putsang, where the villagers greeted us joyfully.

I did not stop at Putsang. Next morning we were off again. We did not march far, but this ensured our reaching Rongyul on the third day from Putsang. There is a pony trail all the way, but it is necessary to cross to the right bank of the Zayul Ngu Chu above the Ata gorge. The ponies were driven into the water, and swam across; we crossed by the rope bridge and camped for the last time opposite the gorge of the Ata Chu. Next day, after a long march, we reached Rongyul. I had not seen ricefields for six months; the rice crop had, of course, been reaped, cattle grazed over the stubble. Tobacco is also grown here. It was like coming into a new world. I stayed the night at the house we had occupied in May, and was warmly welcomed by the headman's family. The old blind headman himself was led in to meet me. He put out his hand, and felt for the sleeve of my jacket. 'Pönpo,' he quavered, 'my people have had a prosperous year, good crops, good health, happiness. Come back again next year, Pönpo.' Could anyone have received a more flattering invitation! However, the headman's dog did not share his master's good opinion of me, and furtively bit me in the leg from behind.

At Rongyul I picked up the luggage in advance, and on the following day we covered the distance to Solé in a few hours. The storm had passed, the sun shone again. Solé, with its warm golden brown rice terraces, its clusters of log houses embowered in trees, and happy-go-lucky serfs, singing over their work in the fields, looked like civilization. It was a gorgeous evening, the great mountain spurs visibly fading into a milky haze far down the valley. I had been marching and climbing for eighteen consecutive days: now I determined to halt for a few days.

CHAPTER XXI

THE GLAMOUR OF THE MOUNTAINS

I HAD a two-fold object in staying at Solé. In the first place I wanted to dry and pack the seeds collected since leaving Modung; in the second place I wished to try and collect seeds of the dwarf Iris which I had failed to get at Modung. I soon satisfied myself that all the Iris capsules were empty. Nevertheless, I believed that with patience I might find the seeds scattered in the earth close to the plants, wherever they grew thickly. And so it proved. On bare dusty slopes facing south I managed to pick up a few hundred seeds. It was slow work, but by devoting two hours a day to it, one in the morning and one in the afternoon, lying full length and going through the dust and debris carefully, my efforts were crowned with success. Such is intensive seed collecting! The day after my arrival the headman of Modung came to see me. He presented me with a short Tibetan sword, or dagger, in a metal sheath, and a chunk of yak meat, so palpably noisome that I had to heave it out of the window. Another visitor was the widow of the late Governor of Zayul. She, however, had not come from Rima just to see me; she was out on a financial tour, collecting the interest on agricultural loans issued by her late husband. In the rather bulky clothes of the rich Tibetan woman, wearing a striped apron, and a little fur-lined cap perched on her head, she looked almost chic. She was a merry widow; though her debtors probably did not think so. She called on me, preceded by her servant bearing two half bricks of Chinese sugar and some walnuts, 'sent with a scarf'. She went about with an armed escort. This orderly possessed a Mauser rifle; for the wily widow had great possessions. However, he regarded his weapon as an encumbrance rather than an asset, and left it behind at my house. Tsumbi was much concerned, no doubt for the widow, and asked me if he should ride after the escort and return the rifle; but I said no, if he needed it he could come back for it. I had already observed that he had no ammunition. The headman of Solé in whose house I stayed was an enormously fat man with a rubicund, even vinous complexion. I never saw him do a hand's turn of work. All day he hung about his veranda, leaning over the rail, telling his beads and muttering prayers. Occasionally he waddled as far as the village shrine, and sat there in the sun, talking to the oldest inhabitant. But he was very rich. He had the dirtiest house, the greediest pigs, the mangiest dogs, the leanest cattle, and the most underfed serfs of any headman in the valley. His serfs, a pleasant looking girl, a crone, and a scarecrow of a man averaged fifty-six to fifty-eight inches high. They were always up before daylight, when a slight frost lay on the ground. Often they were at work by the light of a pine torch at ten o'clock at night. It is no exaggeration to say that they worked eighty hours a week. Their thin clothes were in rags, they slept in kennels. Tired, hungry and cold, they lay down on the wooden floor by a feeble fire, and I could hear them talking and laughing till far into the night. One by one they drop off. The fire goes out. They awake, shivering; dawn is breaking, another day's work begins.

On the 21st we marched a few miles down the valley to Mugu. I had previously ascertained that there are several tracks leading from the Rong Tö valley over the western hills into the headwaters of the Dibang river; but only one of them is now in use. Before—and probably since—the Tibetans began to colonize and to civilize Zayul, the Bebejiya must have made many raids from their scowling jungles into this smiling valley; they would hardly dare do so now. Nevertheless, in a laudable effort to prevent bloodshed, the Tibetan lamas will 'close' a pass, a proceeding which smacks of a more primitive tabu. Over the pass is out of bounds, but a Tibetan hunter who disregards the tabu does so at his own bodily risk rather than spiritual peril. Both sides recognize the justice of this arrangement, and if a Tibetan is slain trespassing in the Bebejiya country, or vice versa, the challenge is not taken up. This makes for peace in a rough and ready way; but of course, the question arises whether the murdered man was trespassing or not. A recognized pass on a recognized route, is a clear enough boundary; when one gets away from the known passes it is more difficult to agree on which side of the boundary you are, particularly if your own version of the affair is no longer available.

I found the people of Mugu willing enough to provide me with coolies and guides, so long as I did not want to cross the pass. Curiously enough, they did not employ the cogent argument against it I have outlined above, but complained of the snow. I easily convinced them that I had no intention of returning to Sadiya by this route by leaving most of my baggage at Mugu. I also demonstrated that there was no great difficulty in crossing the pass in spite of the snow; but that came later. My real object in going, apart from any geographical interest, was to collect seeds of alpine plants; for I knew that the Mishmi hills were enormously rich in flowers of all kinds. The non-expert may wonder what use it was my going up into the snow at the end of November. Obviously there would be no flowers; how then would I know what I was collecting? There is no mystery about it. After years of experience, always be it remembered, in the mountains of S.E. Asia, I have come to know the sort of plants one may expect to find, and what they look like at any season of the year. I know a Gentian, or a Primula, or a Meconopsis, or a Rhododendron, for example, without its flowers as well as a Londoner knows a Plane tree or an Elm without its leaves. All the enlightenment a flower gives me is on the question of colour; admittedly an important factor, but not the only one. As for shrubs, many of them, such as the Barberries, and Cotoneasters, reveal their greatest merit in the autumn or early winter.

    We started up the steep mountainside at noon on November 22nd. Ascending a thousand feet through open Pine forest, we reached a shoulder, and turning the corner began to traverse the precipitous face of a ravine. Gradually we worked our way towards the torrent, climbing down the cliffs, a risky business. When we camped at dusk, we had left the open Pine country, and its burnt undergrowth, and were in thick forest, surrounded by a great variety of stately trees. Almost immediately I met with species of Rhododendron not previously seen in Zayul. All next day we continued to ascend the steep valley through gloomy forests, untroubled by man. Here and there a side stream had cut a slot through the forest screen, affording a glimpse of white sun-drenched cliffs. The air grew colder, the Rhododendrons changed from trees to shrubs, honey golden Larch replaced Spruce, crinkly leafed dwarf Hollies studded with scarlet beads covered the rocks. At night we camped on beds of moss under the elfin wood; the only sound was the reverberating roar of the torrent as it crashed down the long granite stairway on its descent to join the Rong Tö river, now 4000 feet below. I started ahead of the coolies next morning, full of excited expectation, keeping a sharp look out for new shrubs. The ascent was steeper and rockier than ever: patches of snow lay in the hollows; the grim silence of winter deepened. Suddenly we emerged from the heavy forest into the open valley, cold and frozen, but sparkling in the sunshine. Gone was the crashing torrent leaping down the long incline. Instead the stream flowed placidly in a wide flat valley, interrupted by deep glass green pools; a fringe of dwarf Rhododendrons bordered its grassy banks. I could see some distance up the valley, and an array of stark peaks clustered at its head. Wide fans of gravel were spewed down from the gaunt cliffs; and on the rubble grew thousands of alpine plants, all of course, in seed. Beneath the bushes, the easily recognized blackened scapes of *Meconopsis betonicifolia* grew; the capsules were full of seed—which I collected; for who should say that this was not a new variety of the turquoise Tibetan Poppy! Other meadow plants were *Cimicifuga foetens,* and a gentian three feet high.

We camped on the verge of a ponded lake glazed with ice. A flurry of snow enveloped us, but quickly passed. The way up the valley did not look difficult, though the snow-clad ranges which rose on either flank were precipitous. Just opposite our camp a hanging valley hurled a cascade over the cliff; it fell a thousand feet, encased in icicles. Now the Mugu men told me that the way to the pass was not up the main valley, which was the obvious route, but up this appalling cliff into the invisible hanging valley. The threatening clouds drew off, and the stars came out.

I awoke early on November 25th. The thermometer under my tent fly showed 28 degrees of frost; and it was some hours before the sun cleared the ridge and began to lap up the rime which coated the meadow. After breakfast, with four men, I started to climb the 2000 feet to the Jara La. A zigzag path led up through the snow-bound Fir forest to the hanging valley. Reaching the top of the cascade, and looking over the rim of the basin from which it flowed, we saw a beautiful mountain-girt lake. The pass was a thousand feet higher. It was evident that the slopes in summer were a sea of alpine flowers; sticking up through the snow I recognized many. There were thousands of plants of the incomparable gamboge Primula (*P. Agleniana,* var. *atrocrocea),* yellow Poppies (*Meconopsis integrifolia),* and blue Poppies (*M. betonicifolia);* stark and cold as these plants were now, it was not difficult to picture the scene in early summer, when melting snow supplied water, and smoking rain mist drenched the air. There would be creeping Rhododendrons, red as lava glowing through a veil of steam. The last thousand feet up to the top of the ridge, over piles of enormous boulders, and through powdery snow into which we sank knee-deep, was hard labour. I was so exhausted that I seriously asked myself whether it was worth while going to the top, especially as it was quite impossible to see what plants grew here; but I was very anxious to look over into the valley of the Dibang, and virtually into Assam, so I trudged on slowly. At last, after four hours' climbing, we reached the pass. We stood on the Lohit-Dibang divide: below us was a deep valley where the Tangon river had its source. We could see an alpine meadow 2000 feet below, backed by Fir trees. But more prominent were the snow striped razor-edged ridges which thrust themselves down from the north like the fingers of an outspread hand. An icy wind whisked over the pass. We descended a short distance, so as to get a clear view, and sat down for five minutes to recover breath. But in spite of the bright sunshine it was too cold to stay long, and I soon retreated. I estimated the height of the Jara La at about 15,000 feet.

Most of the alpine plants had seeds in their capsules, especially the Gentians, Primulas, Meconopsis, and Nomocharis; only the Rhododendrons had broadcast theirs. These seeds were being frozen hard each night, in stiffened capsules which thawed out under the sun's rays each day. As the frost-coated capsule softens, the seeds become soaked with moisture; and no sooner do they dry in the sun than they are frozen again as the sun drops behind the range. This alternate wetting, drying, and freezing goes on night after night. I may add that all these seeds, collected in November 1933, germinated readily in England in March or April 1934.

On the following day I decided to explore the main valley. It was impossible to follow the stream closely, but we picked up a hunter's path and clambered up and down over the steep rubble chutes, and then over the chaotic boulders of an ancient terminal moraine. The chutes were covered with a thick sweeping growth of stunted Birch; and there were clumps of Silver Fir in the sheltered hollows. The top of the moraine was covered with a deep crust of snow, through which peeped clumps of Rhododendrons.

The valley then broadened out again for the last time, ending in a flat marshy meadow at the foot of a small glacier. There was no exit; the valley narrowed and ended under the cliffs. While strolling on the marsh, I noticed a black bear on the snow slope about a quarter of a mile away. Under our scrutiny he moved slowly up the slope, hugging the fringe of leafless willows, stopping occasionally and sitting down to look at us.

The weather was still perfect, the nights very cold. I had abandoned my Beatrice stove, having no more kerosene left, and I found getting out of bed in the early morning into twenty degrees of frost rather trying. But the cold woke me up while there was yet only a faint light in the East, and the only way to get warm was to kindle what remained of my previous night's fire outside.

I now came to the conclusion that I could not usefully collect any more seeds here, the snow was too deep; and that if I intended to return to Assam by the Delei valley, I had better be going. Another snowstorm might seal all these passes; and although the Dri pass into the Delei valley was quite low (it is said to be under 13,000 feet) it was not likely to be easy. I was inclined to underrate it.

On November 27th therefore we packed up our camp and started down the valley. I proposed to take three days to reach Mugu, the same as on the way up; for although we could easily have done it in two, I wanted plenty of time for seed collecting. On the second day in a gully, the flanks of which were covered with a rabble of high herbs, I suddenly saw the capsules borne stiffly, like squat candles on brackets, of a many-flowered Nomocharis. I told all the coolies to set their loads down and search for more specimens, offering a reward for each one found. We quickly discovered several dozen plants, and collected abundant seed, besides digging up the bulbs. It was impossible to identify the plant with certainty. It might be *Nomocharis pardanthina*, or it might be a new species. But it was the first many-flowered Nomocharis I had seen in Zayul, the others, such as *N. Souliei*, being dwarf plants, not above six inches high and bearing only a solitary flower. Ascending this gully I was finally stopped by a bare granite cliff; and remarkable as it sounds, here at about 10,000 feet altitude on November 27th I saw a flash of mazarin blue and came upon a plant of the Lhagu Gentian, its glazed porcelain cups brimming over with liquid sunshine. When one remembers that every night those cups must be frozen to the consistency of parchment, and that by morning they are brittle with a thick deposit of ice crystals, my astonishment will be understood. The truth is, there are quite a number of what one may call frost-resistant flowers in Tibet. I have already remarked that the alpine flora is necessarily a late one. November and December though cold are the sunniest months in the year, and plants which can put off their flowering till then are certain of ample sunlight. On the other hand they may have to ripen their seeds under snow.

We reached Mugu before midday, and halting only the night, we continued the march to Dri, reached on December 2nd. Perfect weather prevailed though a milky haze made visibility poor. Owing to the great height of the boundary ranges, however, the sun does not long remain in the narrow Rong Tö valley, which cools quickly in the shadows after three o'clock.

The iron red stems of the Pine trees striping the foreground, shot with the haricot red shafts of the setting sun, made a memorable picture. Everywhere the dry undergrowth was burning, and volumes of smoke were being belched into an unclean sky. Many a noble Pine was charred, some past recovery, but there is such a vast extent of forest here, that it could make hardly any difference. In a big ravine below Giwang I noticed a magnificent evergreen tree with a few small cream-coloured flowers. It was 150 feet or more high, with an immense smooth trunk and a fine compact crown of branches like a Dipterocarpus. My attention was first drawn to it by an aromatic scent, not quite familiar, yet recalling some half-forgotten memory. Looking on the ground I saw a wooden fruit twisted and pustular from which peeped a vermilion seed. Then I remembered what the scent was—it was the pungent scent of Magnolia seeds, which are full of oil. This was not the fruit of a Magnolia, however, but of the closely allied genus, Michelia. Continuing the search, I soon picked up some spear-shaped leaves, glistening with long soft silken hairs, and next a few scattered petals. Gradually I matched all the parts to each other, finding small branches blown off with leaves and flowers intact. Lastly I found the tree from which these fragments were derived. I was able to collect a few seeds (*Michelia lanuginosa*).

Below Töyul the valley changes. The mountains begin to recede, the valley widens. There are huddled foothills at the base of the western range. Broad terraces have been carved by streams out of the gravel which once filled the valley; they stand now in ranks, one above the other, cultivated if water can be brought to them. At Dri[1] these high river terraces are conspicuous hundreds of feet above the river.

Many Mishmis had arrived at Dri from the Delei valley. A few had come to buy cattle and were shortly returning, but most of them had come over to work for the Tibetans during the winter.

CHAPTER XXII

OVER THE LAST PASS

BESIDES the ten or twelve coolies needed to carry the loads I wanted with me, and the same number to carry the remaining loads in charge of Tashi by the Lohit route, I required a dozen coolies for carrying rations. I expected that it would take several days to engage so large a number of coolies for the journey to Assam, and actually it took six. Whether I returned to Rima and thence down the Lohit valley the way we had come, or whether I went over to the Dri La into the Delei valley, I had to carry supplies for at least ten days: the country is empty. This increased the number of coolies, thereby making it more difficult. As guide and interpreter I secured Wunju, headman of Giwang, who acted for the Mishmis when they came over into Tibet. Wunju told me he frequently went to Sadiya. As a matter of fact, he had not been there for some years, and was actually 'wanted' there by the police for certain offences. He omitted to tell me this, and on arrival in Sadiya he was recognized and arrested for opium smuggling. He agreed to start on the 7th; by the 12th not only should we be over the pass, but we ought to have reached the first Mishmi village on the other side, called Tajabum. I had visited Tajabum with Hugh Clutterbuck when we ascended the Delei valley in 1928. Meanwhile I fretted at Dri. I felt that the fine weather we were enjoying could not go on for ever; the sooner we got over the pass the better. But the Tibetans were in no hurry; and although the headmen of Rima and of Dri were both actively engaged in recruiting coolies from six villages, it took them several days to engage a sufficient number. Instructions to recruit coolies anxious to go to Sadiya were issued in writing. As the time for leaving approached I got more and more excited and my impatience increased. It was the last leg of the long journey: over the last pass. In a week I should be in Assam, in a fortnight in Sadiya; perhaps I might even arrive in time for Christmas. The eight months' silence would be broken by news and letters from home. Added to all this was the thrill of exploring a new route and of crossing the pass which had baffled

Clutterbuck and myself in 1928. And finally there was the crowning chance of what had been quite a successful season's plant hunting with some dazzling discoveries.

That the atmosphere was much moister here than it was even beyond Putsang, fifty miles further north, was obvious from the heavy dew at night. This soon dried off in the open, but the deep ravines remained moist all day. After several days of poor visibility, the milky haze coagulated into definite clouds, which however remained high. Nevertheless I urged the headmen to hurry up with the arrangements.

At last on December 7th Wunju, arrived from his village bringing four coolies; there was hope that we might start next day, although Wunju declared he would not be ready. The proceedings on the 8th were typical. We rose early, had breakfast and packed. Coolies began to arrive. Then began the long patience. By this time there were twenty or thirty coolies in and about the house, besides a dozen Mishmis looking on. What with them and the loads, and the household trying to get on with their work, the veranda was rather congested. Four men were giving orders at once. The white-haired headman of Rima, who ranks as a minor official when the Governor is away, was nominally in command. His tall bulky figure, dressed in a long greasy sheepskin coat, with long black boots to match, towered over the little coolies. His major-domo, a quiet-spoken man with white hair, my guide Wunju, and another man with a loud voice and a truculent manner, were also issuing orders. Everybody talked at once. Nobody listened, or if they did they didn't obey. About ten o'clock, owing to shortage of coolies, I knocked off one load; an hour later I sacrificed two more, reducing my loads to ten, but increasing the Rima loads by three; we required no less than six additional coolies to carry rations. Soon after eleven Tsumbi told me that everything was ready, so I knew that in an hour or so we really might start. But we didn't. The long and sometimes fierce arguments all had to do with such questions as which headmen were responsible for which coolies, and who was to carry what; there was also a certain amount of civil discord about rations. The headmen fed their own coolies, but of course I paid for all coolies, the money going to their masters. At half-past twelve Tsumbi said for the second time that all was ready, and half an hour later it actually was. We started. However, we did not get far and might just as well have stayed where we were, for after crossing a ravine we camped in the pine forest within shouting distance of the village. Wunju, said there was no water higher up. At nightfall we were hailed, and several men arrived, anxious to go to Sadiya, and a new distribution of coolies and loads took place. Next day we started again. There were twenty of us; sixteen coolies (including a Tibetan woman and two Mishmis), Wunju, Tsumbi, Chimi, and myself. Also a goat. Ascending obliquely up the steep face of the mountain, we entered a ravine far above the torrent. The path was narrow, rocky and dangerous. A false step might hurl one over the precipice, into the stream 2000 feet below; the coolies moved cautiously. A skin of earth, precariously supporting trees, covered the valley wall; how thin it was I realized when at the first sign of water pouring over the cliff, only smooth granite rock showed. We halted on a ledge by the cascade for lunch. Dense thickets of Rhododendrons grew here, including a species I had never seen before. On dissecting a winter bud I found that it had crimson or scarlet flowers. It must be a marvellous sight in full bloom; a plant scarcely five feet high bore over fifty flower buds, each with eight or ten flowers.

That night we camped on the steep forested slope. It was like camping on a gable with trees growing out of it. An enormous granite boulder afforded ample protection. Magnificent trees, Oak and Spruce, Maple and Pine, towered up into a velvet sky, riddled with stars.

Continuing along the hazardous path we climbed more steeply up Brobdingnagian stairs. Suddenly we reached the end of the forest, and looked down on a stony alpine valley not far below; looking ahead, *up* the valley, we saw only snow-clad mountains. Tremendous granite precipices girt us round, the half-frozen torrent tinkled down a stairway encased in ice. From a ledge of the cliff sprang a dwarf Barberry which instantly caught my eye. The solitary flowers, though large, were straw yellow, and quite ordinary, followed by large bluish-violet fruits. The pedicels were long, but stiff: the flowers did not dangle. It was the foliage and the turgid growth which impressed me. The leaves in tufts in the axils of three-pronged spines were polished olive-green above and silver-white beneath; older leaves were turning rich crimson, but retained their high polish. I secured a few seeds of this rock-garden shrub. Another interesting plant here, also in flower but leafless, was a species of Castanopsis. It looked much more like a Hazel; and indeed I put it down as *Corylus ferox*, until I found the sweet chestnut-like prickly fruits actually on the tree. Both these plants grew on the less sheltered side of the gorge, in what may be called Conifer-Rhododendron forest. Above us the terrific precipices were barren, but across the gorge Silver Firs climbed snowwards.

We did not need to descend far to meet the torrent, and we were soon climbing steeply over boulders again. Thickets of large shrubs, including Birch, Rowan, Lilac, Jasmine and Maple lined the slope, but there were patches of open meadow, and clumps of a low bushy Rhododendron (R. *lanatum*)[1] with thick leaves felted beneath with cinnamon fur, as though to keep out the cold; and this it did so effectively that the leaves were not curled up.

We camped under a boulder. The night was brilliant, but towards dawn a hush came over the ice-caked valley, and when I awoke I was horrified to see heavy banks of greasy grey cloud muffling the forest far below the mountain tops. A fine snow was filtering down. After breakfast the clouds disappeared as mysteriously as they had come, and another sunny day followed. It was wasted on us, because we only had to march three miles to the head of the valley, and here, snuggled into the last belt of elfin Birch, we spent another night. It was a most picturesque spot; a frozen lake was tucked away at the foot of a snow-covered wall of mountain. Over 2000 feet above us was the pass, a notch on a razor-backed ridge; the semi-circular wall clasped us to its frozen bosom. Looking back, the level ice-worn valley ended abruptly where the torrent seemed to be sucked noisily down into the bowels of the earth. Far away, framed between the cliff portals, the high peaks across the Rong Tö valley were outlined on a pale blue background of sky.

It was obvious that to-morrow was to be the great day. We had to cross that snow wall and descend to the shelter of the forest on the other side. That meant a long and hard climb. The rock facade, mantled in snow, was implacable. It overshadowed us, calm, aloof, rather contemptuous of the little camp nestling under the bushes by the deep silent lake. It was horribly steep, and the clinging snow had drifted and piled against the fluted cliffs up which we must struggle. If the weather broke during the night, we were defeated on the threshold of success. True the pass would be open again after a spell of fine weather, but we should be driven down to the valley meanwhile, owing to lack of food. Once back in Dri the coolies would disperse to their homes. I could see nothing for it but the dull trudge down the long Lohit valley. On the other hand we had only to scale

ON THE JARA LA: THE ASSAM MOUNTAINS AND SOURCES OF THE DIBANG

this barrier and our troubles would be over; one day between us and the voluptuous Assamese jungle, one hard day's work. Then down, down with the leaping brooks, swelling to eager torrents, swathed in the scents which rise like fragrant mists from the teeming earth and the strange valiant cries of tropic birds: through forest and fen to the level ooze at the foot of the hills. From the top of the pass we should see the grim Tibetan mountains sag swiftly, and the blunt spurs flare away into the dim grey green plains. As we still had most of the day before us, five of the coolies started to carry the rations to a ledge halfway up the mountain, and dump them, so as to lighten the burden for next day. They returned at dusk, reporting the snow very soft. It looked it. Meanwhile a large party of Mishmis had crossed the pass and descended upon us. Great was their astonishment at seeing me. I was hardly less astonished to see bare-legged men—and women too—tramping through the snow; their bodies were warmly clad in short jackets made of Tibetan woollen cloth, but their feet must have been numb. I climbed a scree, and tried to collect Gentian seed; but my eyes kept straying to that formidable barricade, set as a wall to divide Tibetans from Mishmis. The day dragged. I was restless, and walked down the valley. Where the rubble-cluttered floor had been flushed by the torrent, thickets of pygmy Birch grew; their tapering twigs, cobweb grey against the pastel sky, wove intricate tracery. Here and there bunches of rosy pearls drooped from a leaning Rowan. The sun slid swiftly down a long low arc of sky. Most of my coolies spent the afternoon patching their boots and fur caps. The shadows lengthened; and the Gothic spires of Parangkon, a sharp peak we had seen five years previously from far down the Delei valley, threw grotesque melting patterns on the Norman granite walls. The colours of heaven and earth flowed and mixed and set hard again. The cathedral blueness of the sky deepened: a volcanic red throbbed for a time in the west, smouldered, and went out. Immediately the sun disappeared, the great cold crept from its lair, and breathed over the valley; everything withered at its touch. At intervals the lake ice cracked with a ringing echo. By six o'clock the temperature inside my tent had dropped to 22° F.—ten degrees of frost. It was going to be a cold night. I sat by the fire until 9 p.m. toasting my feet, then crept reluctantly into my flea-bag ready dressed except for boots and leather jacket. I slept fitfully. The men had asked to start about 4 a.m.

when a waning moon rose; but I was the first up, and there was no sign of life amongst the coolies at that hour. I hated getting out of my flea-bag into twenty degrees of frost, but it had to be done. Anyhow by evening I consoled myself we would be in Assam. I felt cold, weary and lethargic; no sooner had I put my nose outside the tent than I almost jumped. Clouds were scudding across the moon, which gave no light at all; snow was already falling. We must hurry; now or never over the last hurdle. A wave of warm muggy air had spurted up from the depths of Assam, and was pouring over the pass. I had no breakfast, only a big bowl of hot pemmican. At 7 o'clock we started. My hands and feet were already numb. Skirting the frozen lake, we crossed the steep valley which fed it, and started up a wide torrent bed, lined with rigid bushes of Rhododendron. The water had frozen, encasing the stones in a film of ice over which lay a quilt of snow. Time and again we slipped on the ice slope, then hugged the bushes, preferring to be whipped across the face by a hard branch rather than break a limb. At last we reached the top of the fall. The slope eased off, and we trudged across an open snowfield to the foot of a couloir. The bulk of the mountain rose directly above us, buttress on buttress. The snow was piled into the couloir and plastered in fleecy layers against the cliff. At every step we sank in almost to our knees. Steeper and steeper grew the ascent, deeper and deeper the snow. I could not go fast enough to get warm. In two hours we reached the ration dump, and ten minutes was spent rearranging the loads. Then we plodded on, now sinking to our knees in the snow and tripping over unseen tuffets of Rhododendron. Another rest enabled me to collect some Allium seed where the snow had partly melted. From here to the top the mountain was completely exposed; it had caught the full fury of the last storm, and the snow had ironed out all the unevenness. The Mishmis, who had crossed the previous day, had bequeathed to us a trail; yet we found it easier to make one for ourselves than to follow theirs. In places the rocks were so steep we could only clamber up them with mutual assistance. Several times I stepped into a drift and floundered waist deep in the snow. We had now been climbing five and a half hours. Close above was the Dri Pass, a knife-edged ridge, with the wind driving the snow in our faces. Panting, I scrambled up the last few yards. At last! Assam lay right at my feet; Tibet lay behind. The Dri Pass crossed, another ambition fulfilled; but I was too exhausted to cheer.

That first view of Assam was more inspiring than reassuring. Nothing but a succession of wedge-shaped ridges, striped and dappled with snow, and the flocculent clouds lying like loose packing between them. Now and then a shard of sunlight pierced the clouds, and lighted up a more distant ridge. Far below, but hidden by the slope of the mountain, was an abyss, the actual source of the Delei river. Picking our way carefully over piles of slippery slabs, we presently got off the snow which on the warmer Indian side did not extend far. Then began a long scramble towards the valley. At three o'clock we came to a trickle of water, and throwing ourselves down out of the wind enjoyed a much needed meal. I collected some alpine seeds here, notably a Cyananthus. At this point shrubs began to appear, at first small and scattered, but presently gaining confidence and strength to form a dense scrub. *Rhododendron lanatum* and a form of *R. cerasinum* were prominent. The latter has scalding crimson flowers with a ring of five large circular glossy coal-black honey glands at the base of the corolla. We had hardly reached the outposts of the forest than it began to snow violently. By the time we reached the valley it was dark; but the deepening snow threw up enough light to enable us to make some sort of camp. We put up the tents: but we could not get the fires going properly, having no dry firewood. Finally we all threw ourselves down to sleep, too exhausted to eat. The crisis was over; reaction after the excitement now set in. I could not sleep.

CHAPTER XXIII

THE MISHMI HILLS

Snow fell steadily all night. 'So this is Assam!' I said coldly to myself as I got up. The whole valley was under snow, a wet blanket of cloud hung low over the mountains, and it was snowing when, feeling rather miserable, we started down the valley. The Delei river was a brook lined with bushes. It was difficult to recognize any plants even in the open meadow, but I noticed the bleached fruits of an Onion (Allium), St. John's Wort, Omphalogramma (probably *O. Souliei*), and Nomocharis. A low growing Rhododendron with silver-plated leaves like *R. sanguineum* showed here and there through a fleece of snow, and in the Fir forest were scattered plants of *Primula geranioides*. But it was a poor day for the collector; the collector himself was feeling poorly. Towards midday the hard granular snow which had been falling all night gave place to large soft flakes which began to melt as they fell. After marching a few miles and descending scarcely at all we came to some rocks in the forest. Everyone was exhausted. It had stopped snowing and only a film of melting snow covered the ground. But the sky looked threatening, and if more snow fell in the night, the coolies would be in a bad way. There was good shelter here under the rocks, and plenty of daylight to light fires and cook hot food. So we camped. Towards evening the sky began to clear. Three shivering ill-clad Mishmis who had intended to cross the pass into Tibet, but had been driven back by the snow, joined us. They were returning to their village to await a more favourable opportunity. They told us it was impossible to forecast a storm. The bad weather might last a day or two or it might last a week. As it turned out, the next day, December 14th, was fine, and we met many Mishmis coming up.

My boots were frozen stiff in my tent that night; and the coolies being sluggish, we started late. There was an ominous air of disaffection abroad. We followed the river closely, and crossed two big streams which in the absence of bridges, would be quite impassable during the rainy season.[1] I had a hazy idea that we should reach a part of the valley familiar to me this day, and the Mishmi village of Tajabum next day. But Wunju, quickly cured my optimism: we should not, he said, reach Tajabum next day, or the day after either. I rated him for misleading me about the distance and he went off sulkily to his camp, with his men, two of whom spoke Mishmi. Later they were joined by a party of Mishmis on their way to the pass. They sat round their fire talking in low tones till a late hour. Trouble was brewing.

Next morning matters came to a head. Most of the coolies, my two servants and myself were camped close together; Wunju's, party, consisting of himself, one Mishmi and five Tibetans, were by themselves. When we were ready to start we called to Wunju, who answered but did not come. At last Tsumbi went off to bring him: and Wunju came with an ill grace. Also his coolies. They had a grievance. They did not want to go on. I must raise their pay. Now the arrangement about pay had been settled with the headmen at Dri, before starting, and it was on a generous scale. The headmen had, in fact, dictated the rate of pay to me, and I had accepted it without demur. Wunju, by organizing a mutiny on the high lands, hoped to paralyse me. I was furious. Nothing whatever would make me give way. At this point two of the five Tibetans with Wunju, dissociated themselves from the mutineers. They told Tsumbi that they had not come when he called them because they had not finished breakfast. They were satisfied with their pay, and had agreed to it, they did not want it altered—though of course, if I liked to give them an extra present when the job was finished it would be appreciated. That sadly reduced the mutineers; there remained only Wunju, a Mishmi, and three Tibetans. We were all gathered there in the forest, the loads tied up, the fires dying down, the loyal party anxious to start. A good deal depended on the next few minutes. Wunju, and his mutineers stood apart, looking sullen; the two waverers joined the loyal party, and I called Tsumbi to my side. 'Tell him,' I said, pointing to Wunju, 'that we are starting now. He can go back the way he came. I will jettison three loads, comprising a tent, half the rations and stores, some botanical drying paper, and a few other things. The fourth load will be distributed amongst the other coolies.' Some of the loyal coolies laughed; Wunju, looked crestfallen. 'Of course,' I added, as though it were an afterthought, 'those who go back now will get no pay, and we want the rations.' Ten minutes later we were all marching down the valley again.

We met many Mishmis on their way up to Tibet; old men and women, girls and a few young bucks. Every other man had a cap gun. One I looked at was by Hollis & Son, the name just decipherable; it was at least fifty years old, but looked a serviceable weapon.

We did a long day's march, descending probably a thousand feet, and camping in temperate evergreen rain forest. The trees were swathed in moss, and there were numerous epiphytic Rhododendrons, one with small yellow flowers; also a scarlet flowered Aeschynanthus. Earlier in the day we had passed through a grove of magnificent Magnolias (*M. Campbellii*). The Delei river was now a formidable torrent, roaring down a steep bed amongst enormous blocks of stone. The steep sides of the gorge were lined with Rhododendrons. Some Mishmis were camped here, digging up the roots of a medicinal plant, *Coptis Teeta*. Clearings are made in the forest, and the plant spreads rapidly of its own accord.

It was now nine days since we had started from Dri, and still Wunju, insisted we could not reach Tajabum the next day. In an effort to do so, we marched till after dark, all to no purpose. The Delei river was now plunging rapidly down into its gorge, and the path climbed steeply up the mountainside, but distances were greater than they looked, as we wound our way over spurs and into deep gullies and out again. The peak called Kaso came into view; its flanks were deep under snow, and all the beautiful alpine Primulas we had found there five years previously were buried; but it was good to catch a glimpse of a familiar peak, even of one so remote and implacable as Kaso.

We camped once more, this time high above the river in gnarled forest. I did not realize how close we were to the farthest point we had reached in 1928. Starting ahead of the coolies next day, I climbed a few hundred feet and almost immediately emerged from the forest on to an open scrub-clad cliff. A narrow path traversed the face, and I went on to the top. Nearly 3000 feet below flowed the Delei river, white with foam; but not a sound came up to me. Ahead, the steep mountains began to open out, and unroll in a wider panorama; the spurs melted one by one into the milky haze. Beyond the last distant ridge lay the plains. It was the tenth day since leaving Dri; the village of Tajabum lay below.

We halted on the ridge above Cha Che, where I had sat on that wet May day of 1928, thinking that the pass was but two or three easy days' march away. I knew better now. Then we descended the terrific cliff which leads down to the big torrent, past the little rock shelf cleared by the inquiring Chinaman who came over from Dri to have a look at Assam; he had looked at it—from Cha Che, and gone back whence he came. Clutterbuck and I had spent a memorable night on this ledge. Then the future had seemed black, because we had been unable to reach the pass. How different the outlook now, the pass crossed, the weather gloriously fine, and in five days—well, a week—perhaps ten days if things did not go well, I should be in Sadiya. But our adventures and anxieties were not over yet. Passing Tajabum—a landmark rather than a village, for there are only two huts, and the occupants had gone to Tibet—in the afternoon we found ourselves on a narrow path high above the river, but descending to it; we had to cross to the other bank farther on. Presently the flimsy bamboo native bridge came into view half a mile away. We were still some height above it, looking down on to it, when suddenly Tsumbi seized my arm. 'Look, *sahib*!' he said, and pointed to the bridge.

I now saw three Mishmis, with their knives drawn, hacking feverishly at the bridge. They had lain in hiding until we came into view, and were now bent on destroying the bridge to prevent our crossing. We shouted, and started to run; two coolies, who with Chimi, had gone ahead unknown to the Mishmis, were now close to the bridge. No sooner did they appear on the river bank than the gallant defenders fled: as we rushed down the path we saw them running up the hill on the opposite side: soon they disappeared into the jungle. A few minutes later we saw Chimi and the two coolies working their way gingerly across the partly destroyed bridge. We had surprised the wreckers in the nick of time; the bridge was passable—but only just, and not for a coolie carrying a load. We tied up a few of the loose bamboos, and got the whole party across in an hour. Then we camped on a piece of cultivated ground. There were scattered huts higher up, but we could not see them.

I felt rather uneasy. By now all the Mishmis in the Delei valley must have known I was coming; and they did not appear to be friendly. I called Wunju,.

'Why did you let the Mishmis cut the bridge? Do they want to stop me from reaching Sadiya? They are your friends, you had better go ahead and tell them that I must be allowed to pass unmolested.'

But Wunju, only smiled enigmatically. 'They are quite friendly, Pönpo, and they will not stop you. They tried to cut the bridge because the last time two Englishmen were here, they paid a good price to have a bridge put up. They thought you would do the same.' It was evident that Wunju was in the plot. Assuredly Clutterbuck and I had made our mark last time we were here, to be so affectionately remembered.

There was a scene next morning. One of the Mishmis, he of the mutineer party, wished to return to his village right away. As that would have meant leaving his load behind, I said he must go on with us till we could get another coolie. He refused. I said very well, I would not pay him till he did; whereupon he drew his knife, waved it dramatically, pranced about lopping down plants (imaginary enemies), and went off in a huff. We then set about re-arranging the loads, being one coolie under strength. After a while the truant came back. He grinned at me, rather sourly though, and began to tie up his load. Someone had taken a battered tin I had given him, and he tried to recover it; but to his disgust the new owner refused to surrender it. At last we started on the most difficult and tiring march of all: a traverse round the cliff, then up notched logs, hand over hand, holding on to creepers and roots. Exhausted, we reached the village towards dusk, and created a great deal of excitement. Was it a Tibetan invasion? Fifty men, women and children watched us, suspiciously. No headman came, with friendly overtures, to offer me fowls or eggs; and we had to pay heavily for our firewood.

Wunju, of the winning voice asked me to rest here a day, as he wanted to sell salt. I agreed to go only as far as Minutang, where we had made our base camp in 1928: a large village just across the next stream; and there we spent the day, while a concourse of Mishmis feasted their eyes on us. It surprised me to see how prosperous and neat the village was, how clean the fields, the corn cobs stored within bamboo fences, roofed with grass, the paths trim. The truculent *gam* of Peti greeted me, stared hard, apparently recognized me, and shouted a question. He had not altered much in five years.

I was glad to get off next day. Then followed three tiresome marches up and down the water-grooved forested spurs which flare out to the Delei river. Now we were getting down to the hill jungle. Screw pines (Pandanus) and Palms began to appear. One night darkness caught us on the narrow path in dense forest. There was no water. When at last we found a stream there was no place to camp. We lay down under the trees anyhow, and at daylight resumed our march.

Here and there a long Mishmi hut appeared on the slope above us. On the down-hill side is a veranda, hidden by a tall bamboo fence up which runner beans grow. The effect is decidedly picturesque, and the object of the screen is laudable, for the veranda is the family latrine. Each room has its exit on to the veranda which is partitioned.

One hut was surrounded by a fence of spiked bamboo, entered by a lift-up door which fairly bristled with sharp bamboo spears. This was not to keep out robbers, but to keep out evil spirits: there was an epidemic in the village. I heard that the headmen were wrath with Wunju, for bringing me into the Delei valley. They hold that any traveller wishing to pass through their territory should pay a tax, and believe they have the right to levy such transit duty; but the bridge ruse having failed, they had no method of enforcing it on me. The truth is the Mishmis have never been given a lesson in manners, and have quite an exaggerated idea of their own importance. They talk a great deal, but their deeds are few; and they have yet to learn that the only title deeds to land is prowess in arms. They are just a very dirty jungle tribe, and no sophistry can make them out to be anything more.

We passed through Chibaon on the 23rd and left the hills altogether, descending at last into the river bed, and camping on a sandbank. In the middle of the night I awoke suddenly at the sound of a cry, 'Stones! stones!' The Tibetans were rushing away from the cliff beneath which they had been sleeping, and I heard the clatter of falling stones as some animal crept across the face high up. Gradually we got to sleep again. It was a beautiful night, warm compared with the mountains; yet everything was drenched with dew in the morning. The march was a difficult one—but it was the last. Where the river skids across to the cliff there is no path. You haul yourself up the cliff by means of cane ropes, straddle from ledge to ledge, a hundred feet above the river, and descend to the beach again. Progress was slow, but when we camped, like Moses, in the rushes, the worst was over. At noon on Christmas day, the old suspension bridge on the Lohit Valley Road, which we had crossed ten months previously, came into view, and we halted for lunch. What a relief it was to strike the main road again, to walk upright, to have one's hands free! It was warm too: the Tibetans, dressed for a cold climate, were casting off some of their garments.

I heard the familiar cry of the coppersmith, and the chatter of babblers. A cormorant flew down the valley. Only a few miles more. We marched for eight hours on Boxing Day, halting only when it was dark. It had rained in the night, and the jungle was dripping; but it cooled the air. We passed a number of Mishmis, and they halted and spoke politely to us, in Assamese. They were returning from Sadiya. I learnt that Tashi and his gang had not yet passed; also that the Political Officer had sent my English mail and some bullion to meet me in Rima! It was annoying to think that after all, when I reached Sadiya, there would still be no letters for me!

December 27th. Last day! Starting early, we reached the Tidding river at noon, and sat down on the hot dry sand for lunch. Then on once more with a final spurt, the excitement growing.

We reached the suspension bridge over the Tidding river, and the Tibetans, who had never seen such a thing in their lives before, after stepping gingerly on to it, bounded and danced across the swaying structure, laughing like children. Then they peered into the iron trellis work pillars, fingered the wire cables, examined the planks, and thoroughly approved of the whole thing. Jauntily we marched the last mile, striding out now, till suddenly we saw the little white bungalow of Theronliang.

'Is this Sadiya?' asked one of the Tibetans innocently.

It was the third time I had marched safely out of the Mishmi Hills. Theronliang is always something of an anticlimax; one is faced with another climb just when one's troubles seem over.

Next morning we crossed the Tidding ridge at 6000 feet. Looking back up the gorge of the Lohit at the jigsaw mountains, I took a last look at the snow; in front, at my feet, stretched the plains swaddled in mist. With fast-beating heart I strode down the path, not even halting at Dreyi for lunch. I ate a piece of chocolate instead. It was late on a golden afternoon when I swung into Denning. I was tired after my twenty-one mile walk. The Political Officer, who had motored in from Sadiya after lunch, greeted me warmly: it was nearly six months since I had seen a white man. But he would not hear of my sleeping the night at Denning, as I had intended to do; nothing would satisfy him but that we start at once on our fifty-mile drive through the jungle to Sadiya.

It was dusk when we started, but the evening was gorgeous. A full moon rose behind us over the mountains as we drove westwards into the red embers of sunset; night came like a narcotic to my jangling nerves. The headlights clove a path through the jungle, but intensified the darkness beyond. Approaching Sadiya, belts of pearl white mist lay across the road and engulfed us.

At ten o'clock we reached the Political Officer's bungalow, where I was made welcome. It was many months since I had covered 70 miles in one day. It usually took a week at least. I had a hot bath…

NOTES ON CHAPTER I

1. It is not always realized by the general public what a large proportion of our trees, shrubs and garden flowers, even the commonest, have been introduced from abroad. Altogether many thousands, including the Horse Chestnut and the Plane tree, Wall-flower, Crocus, Tulip, Chrysanthemum, Michaelmas Daisy and Dahlia have come to us from overseas. Amongst those which I have myself discovered and introduced into the British Isles, may be mentioned: *Meconopsis betonicifolia* (*'M. Baileyi'* of catalogues, the Tibetan blue poppy); *Cyananthus lobatus, Primula Florindae, P. microdonta, Lilium Wardii* and *Rhododendron leucaspis,* from Tibet; *Rhododendron Wardii, R. lysolepis, R. melinanthum, Gentiana trichotoma, Primula chungensis, P. Bulleyana* and *Campanula calcicola,* from China; *Cypripedium Wardii, Rhododendron imperator, Gentiana gilvostriata, Meconopsis violacea* and *Berberis hypokerina,* from Burma; *Leycesteria crocothyrsos, Rhododendron crebreflorum, R. riparium, Primula Normaniana* and *P. concholoba,* from Assam.
2. The Mishmi Tribes—Digaru, Bebejiya, Taroan, Chulikata and Miju, inhabit the Assam frontier hills which enclose the basins of the Dibang and Lohit rivers with their tributaries, known as the Mishmi hills. The tribes speak different languages, and follow different customs; but they are all probably of Tibetan origin, and entered their present home from south-eastern Tibet. Their country is extraordinarily precipitous, covered with dense forests, and almost impenetrable. It covers a superficial area of between 6000 and 8000 square miles. The highest peaks, on the Tibetan frontier, attain 19,000 feet. The whole hill tract is a botanical paradise, and has been very little explored. Endemic plants known include *Primula rubra, P. Normaniana,* and *P. Clutterbuckii, Rhododendron mishmiense* and *R. patulum, Leycesteria crocothyrsos, Molineria oligantha, Gaultheria codonantha* and others.
3. Denning has a rainfall which varies from about 230 to over 300 inches. Most of it falls between June and September. In July, 1931, 73 inches of rain fell. The record, since measurements were first taken, is held by July, 1927, with 76 inches, *two feet* falling in 24 hours on July 7th. Whenever the rainfall at Denning exceeds five inches in 24 hours, the news is wired to Poona, and warnings are sent all over India. Such a rainfall invariably means floods lower down the Brahmaputra and inundated paddy fields. In 1932 the total rainfall was 232 inches. No wonder the skin of the mountains peels off like a transfer! Tremendous landslips scar the outer hills; these block the rivers and cause great havoc when they give way.

NOTES

4. In the winter of 1911-12, and again in 1912-13, the Indian Government sent an expedition up the Lohit valley to survey the country and to fix the frontier. A mule road was built from Denning to a village called Walong, a few miles south of Rima. It was intended to establish one or more posts in the Lohit valley, and to administer the hill tracts. Owing to the war, the project was abandoned, and within a few years the road was swallowed up in the jungle. Much of it, though overgrown and not used by the Mishmis, can still be traced, nor would it cost very much to reopen it for mule traffic.

NOTES ON CHAPTER II

1. The Lohit valley is filled with a type of forest known to botanists as Indo-Malayan, which covers much of south-eastern Asia. The vegetation, up to about 3000 feet, is sub-tropical, and owing to the lack of any marked dry season or cold winter, the forest is mainly evergreen. At higher levels however, even quite close to the plains, deciduous temperate forest, not unlike that met with in England, begins. This is the zone of Oaks, Maples, Hollies, Birch, Hornbeam; but there are also Magnolias, Rhododendrons, Laurels and other trees not found in England. Still higher, a third zone of forest, composed mainly of Coniferous trees and Rhododendrons, is found; and where the hills exceed 13,000 feet, a fourth zone, of Silver Fir. Thus it is possible, without travelling many miles from the sub-tropical jungle, to pass through forests of European and finally of sub-arctic appearance. Continuing up the Lohit valley, near Rima, at an altitude of 4000 feet, the vegetation changes its character, chiefly owing to the fact that the rainfall is less, the atmosphere drier, and the winter colder. Pine forest replaces Indo-Malayan forest. This Pine forest occurs between 4000 and 6000 feet, above which it is replaced by mixed temperate forest. It will be readily understood that in the course of the journey, I passed through many zones of vegetation and several distinct plant regions.
2. Rima is the name marked on all maps of Tibet. Actually the village where we stayed, and where the Governor resides, is called Shigatang. It is situated about a mile north of Rima, which name might conveniently be applied to the whole cultivated district below the confluence of the two rivers. I have so used it here. There are seven villages altogether. I estimated the number of houses at eighty (about half of them in Rima and Shigatang) and the total population at between 350 and 400. There is one Mishmi village.

NOTES

3. 'Zayul', conventional spelling for Tsayul—the 'hot country'. It comprises the two branches of the Lohit river or Zayul Chu, with their tributaries, and covers a superficial area of about 4500 square miles with an estimated population of between 2000 and 3000 persons, the great majority of whom live in the two main river valleys below 8000 feet.
4. The eastern branch has always been regarded by European geographers as the larger, and called the Zayul Chu; the western branch is regarded as a tributary, under the name Rong Tö Chu (Rong Thod Chu on maps). The Tibetans say that the two branches are equal. The combined river is called the Rong me ('lower river' or 'valley') in Rima. It seems likely that in summer, at any rate, the western branch discharges more water than the eastern branch.

NOTES ON CHAPTER III

1   A crested Iris related to *I. Wattii*. In cultivation from seeds which I gathered in Upper Burma in 1931. Tibet is not rich in Irises, though one might expect to find them in the drier parts. Probably the soil is too cold, and the winters too long—though this would not apply to the deep river gorges. Most of the Irises known come from the wetter forested parts of Tibet. They include *I. kumaonensis*, '*I. Wattii*' (above-mentioned), *I. Delavayi*, from Zayul, and another species from the Tsangpo gorge. On the other plateau occur *I. Clarkei* and *I. goniocarpa*. Another dwarf species is found at Shugden Gompa.
2   The principal villages up the Rong Tö valley and the stages are as follows: Sachung; Dri; Töyul; Giwang; Mugu; Solé; Rongyul; (camp); Putsang. Or, turning up the Ata Chu, towards the pass, after camping one night above Rongyul, Modung; Ata. There are a few villages on the left bank also; but rope bridges are scarce, though during the winter rafts can cross the river at many points.

NOTE ON CHAPTER VI

1. The temple was dedicated to the spirits of this romantic spot by the late Kalon Lama, who was General Secretary to the Tibetan Government. This Kalon Lama commanded the Tibetan troops against the Chinese, first in Lhasa, later on the eastern frontier. In 1913 he was sent to Chamdo to organize the Kampas in their resistance to a Chinese advance; and at Chamdo he died.

NOTES

When dedicating the temple of Gechi he chanted:
Nam tu sum,
Ri tu sum,
Chu tu sum,

the literal meaning of which is 'sky triangle, mountain triangle, river triangle', in allusion to the meeting of the three rivers. In Tibet every lake, river and mountain is presided over by a spirit; and where three spirits come together, as at a river junction such as this, their power is very great. The three spirits here were invoked to guard, protect and bless Gechi Gompa.

NOTES ON CHAPTER IX

1. Neither A. K. nor I was able to measure the height of the Ata Kang La. Glacier Camp, however, was 13,813 feet, and the foot of the glacier on the north side of the pass was 14,367 feet. The estimated height of the pass therefore, 16,000 feet, is probably not far wrong.
2. Vegetation soon became general and included dwarf Rhododendrons on the rocks, mostly the same species as those already met with at Chutong, and a great variety of alpine flowers growing amongst low bushes of willow. Along the river bank and in the pastures, besides the flowers mentioned, where Draba, Saussurea, *Primula alta,* Fritillaria, Allium, Nomocharis, Geranium and many more.
3. Crossing the Ata Kang La was exactly like crossing the Himalaya by the Tang La on the main road from Sikkim to Tibet. To the south was a forested land, with plenty of cultivation; to the north, an alpine plateau land without forest, a grazing land with little cultivation. The plateau flora belongs to the Central Asiatic region of botanists. Characteristic of this region are species of Oxytropis and Astragalus.

NOTES ON CHAPTER X

1. Between thirty and forty species of shrubs are found in the vicinity of Shugden Gompa. The only trees are *Picea lichiangensis,* and Poplar, the last quite small.
2. The rocks of SE. Tibet, which I observed, appear to be very similar to those described from NW. Yunnan by the late Professor J. W. Gregory. The Yunnan mountains belong to the Indo-Malayan system, which was uplifted long before the uplift of the Alpine Himalayan system. According to the

prevailing view, this older land mass obstructed the Himalayan uplift, and diverted it southwards. According to the view put forward by Kropotkin, and again by Gregory, and here reinforced and emphasized, the Indo-Malayan system by no means blocked the Himalayan movement, which affected the older rocks to such an extent as entirely to alter the structure of the country.

3. A species of the 'Nivales' section, similar to *P. melanops*, or *P. calliantha*. Of eight species of Primula found in the neighbourhood of Shugden Gompa, four belonged to the 'Nivales' section and one each to the sections Denticulata, Sikkimenses, Bullatae and Farinosae. It would seem then that the Nivales are the typical Primulas of the upper gorge country, and of the outer plateau, that is of the high and comparatively dry country.

NOTE ON CHAPTER XI

1. Formerly Shugden Gompa was under Sangachu Dzong, and therefore part of Zayul. Some years ago, however, the people petitioned to be separated from Zayul and the request was granted. There are, however, villages north of the great range, and even in the Salween valley, which are under Sangachu Dzong. The Tibetans have little regard for what we should consider natural geographical boundaries: hence the difficulty sometimes of selecting a frontier.

NOTES ON CHAPTER XIII

1. The average fall of the Tsangpo between the plateau and the plains is therefore thirty feet a mile; but for the first fifty miles below Pe it is much more than this. The average fall of the Salween is only 10 feet a mile; though it probably exceeds this in its gorge further south. Thus the Salween is nearer its base level; unless the upper course of the Tsangpo has recently been uplifted, the Salween must be the older river.

The Salween is known to the Tibetans as Gyama (or Gyamo) Nsru Chu, and also as Dza Chu or Dza Khog Ngu Chu. The word 'ngu' occurs in connection with several Tibetan rivers, e.g. the upper Rong Tö Chu is called Zayul Ngu Chu by the people of Putsang. 'Ngu' means literally 'sweat', and the explanation of the name, according to some Tibetans, is that the river is sweated out of the earth ('dza' means earth or clay), or that water exudes everywhere from the earth, as it would appear to do in a country of

NOTES

clay hills. This of course could only apply to the extreme headwaters of the Salween, not to the deep arid valley where I saw it; even so, it does not quite tally with Rockhill's account. (See *A Journey Through Mongolia and Tibet*, by W. W. Rockhill.) Nor does it apply to the sources of the Rong Tö, which descend from glaciers.
2. Two lizards noticed in the Salween valley are definitely Chinese, and a toad common there is Palaearctic. The butterflies are mainly Chinese. The flora is rather specialized, but its affinities are with the Chinese (Eastern Asiatic) flora. The following are a few of the plants found in the arid valleys: *Androsace coccinea, Amphicome arguta, Eremurus chinensis, Saxifraga candelabrum, Dregea sinensis, Cheilanthes farinosa, Bauhinia densiflora*, besides those already mentioned.
3. Primulas of the section Bullatae were formerly thought to be confined to Western China. This is the farthest western record of any species of the section.

NOTE ON CHAPTER XIV

1. The direct route from Chamdo to Shugden Gompa is down the Mekong, via Dzo, Taka and Wayo; thence via Yushi, over a high pass, to the Yü Chu, and southwards again through Bomda Gompa, and Tento Gompa; thence over another high pass to Yata on the Salween and south again through Roaba, Wobo and Osetang; finally over a third high pass to Sangachu Dzong, passing through the villages of Trongyü, Tsudru and Giying; whence Shugden Gompa is reached in two days. The journey normally takes about twenty-four days, but could of course be done in less. There are other routes by which Chamdo can be reached. All this part of south-eastern Tibet is well supplied with pack 'roads' of a sort, and the country is comparatively thickly populated. The obstacles to travel in any given direction are the high ranges of rocky mountains and the deep river gorges between them. Roads east and west therefore are conditioned by passes over the ranges and the bridges over the rivers. For an account of this region, see that excellent book, *Travels of a Consular Officer in Eastern Tibet*, by Eric Teichman.

NOTES ON CHAPTER XV

1. Zo La = Bailey's Jo La.
2. On the north side of the pass the alpine flora was similar to that seen

NOTES

on the Traki La. Besides *Gentiana trichotoma*—which does not appear to extend north of Shugden Gompa—I noticed: Polygonum (P. Forrestii and others), Pedicularis (several species), Geranium, Swertia, Dracocephalums, Cremanthodiums, Asters, Crepis, Lactuca, Saxifrages, dwarf Aconite, Arenaria, Astragalus and Cyananthus, all in bloom; *Primula minor* in fruit. This is rather different from the alpine flora of the Ata Kang La, and suggests that the Zo La has a drier climate: a conclusion supported by the fact that one can cross it at any season.

3. The woody vegetation was a very pale reflection of the forests in the Rong Tö valley. Besides the shrubs mentioned, I noticed all those which grew at Shugden Gompa. Three Conifers, *Pinus excelsa,* Picea and Abies, ascend to the tree line at about 13,000 feet, though they only occur scattered. Two species of Birch are mixed up with Rhododendrons and Conifers at this altitude.

NOTE ON CHAPTER XVII

1. According to the Abbot of Shugden Gompa, the three most sacred mountains in Tibet are Ka-Kar-Po, on the Yunnan border, Lho-Tsa-Ri, south of the Tsangpo, south-east of Lhasa, and Tö Kang-Ri, meaning thereby a peak in upper Tibet, towards Ladak, and therefore almost certainly Kailas. The usual Tibetan name for Kailas is however Kang Rim-Po-Che, and since there are other sacred mountains in Tibet known as Tö Kang-Ri, it is possible that one of these near Mount Everest is meant. But the fame of Kailas is so great that it is much the most likely.

Ka-Kar-Po, a peak about 23,000 feet high, stands between the Salween and Mekong rivers, in latitude 28° 30' N. Lho-Tsa-Ri is a forest-covered peak difficult of access. Even the pilgrims are said to need local guides; they frequently perish from the attacks of wild beasts. On modern maps of Tibet Tsari is no longer shown as a mountain, but as a district. The peak called by Bailey Takpa Shiri is probably identical with Tsa-Ri. (See F. M. Bailey, *Geographical Journal,* October, 1914.)

NOTE ON CHAPTER XX

1. This is the pilgrims' path. The priests of Shinki Gompa appear to be the most faithful pilgrims. Starting from Shinki in the summer, they descend the Chimdru river, almost to its junction with the Tsangpo, continue up

NOTES

the left bank of the latter and cross the Sü La, into the valley of the Nagong Chu. Thence they continue up the Nagong Chu to Shugden Gompa, cross the Zo La, and passing by Sangachu Dzong reach Rima. In the following summer they return up the Rong Tö valley to Putsang, and crossing the Kangri Karpo La, arrive back where they started. Nominally the journey takes about three months; actually it takes a year. Chömbö is only one of several sacred peaks in eastern Tibet. Another is Ka-kar-po, on the Chinese frontier; Pilgrims march round this peak also.

NOTE ON CHAPTER XXI

1. Dri is a Tibetan village by the Rong Tö Chu. The pass over the western range into the Delei valley is therefore called the Dri La. This must not be confused with the headwater stream of the Dibang, which is called the Dri river. The Dri river rises near the Kangri Karpo La, and when I was near that pass in November, I was separated from the Dri river only by a single range. On the Survey of India map, the Dri La is called Glei Dakhru, a name obtained from the Mishmis; it was from that side the pass was seen.

NOTE ON CHAPTER XXII

1. *Rhododendron lanatum* is a Himalayan species and was not known to occur outside Sikkim until Clutterbuck and I discovered it in the Mishmi hills, 450 miles to the east, in 1928. The Sikkim plant has long been in cultivation, but it grows very slowly and does not thrive. The Mishmi hills dwarf variety is also in cultivation, but has not flowered as yet; it is only six years old.

NOTE ON CHAPTER XXIII

1. The Delei valley was surveyed in 1912-13, but the surveyors were unable to reach the head of the valley, and, deceived by appearances, assumed too much. Thus the map shows only two big tributaries on the right bank, rising from the main divide, instead of four.

INDEX

I
PLANTS

Abies, 61, 120, 199
Abies Delavayi, xv
Abies Pindrow, xv
Abies Webbiana, 62
Acer, xv
Acer stachyophyllum, 54
Acer Wardii, 34
Aconite (dwarf), 199
Adenophora, 128
Aeschynanthus, 187
Ailanthus, 29, 157
Ajuga ovalifolia, 90
Allium, xv, 183, 185, 196
Alnus nepalensis, 9
Amphicome arguta, 198
Androsace, 110, 128
Androsace Chamaejasme, 74, 112
Androsace coccinea, 198
Anemone, 64, 74, 81, 96, 112, 125
Anemone rupicola, 81
Arenaria, 199
Aristolochia Griffithii, 52
Artemisia, 16
Arum (Arisaema), 53, 164
Ash (Indian), 56
Aster, xv, 74, 76, 81, 89, 90, 112, 116, 118, 121, 140, 199
Astragalus, xv, 74, 76, 79, 196
Bamboo, 164, 167
Bamboo grass, 34
Barberry (dwarf), 179
Bauhinia densiflora, 198
Berberis, 75, 157
Berberis hypokerina, 193
Birch, xv, 29, 53, 120, 128, 134, 173, 180, 182, 194, 199
Buckthorn, 61, 99
Buddleia, 107
Caltha, 112

Campanula calcicola, 193
Caragana jubata, 75
Carmine Cherry, 18
Caryopteris, xvi, 107
Cassiope, 75
Castanopsis, 179
Caulophyllum robustum, 52
Ceratostigma, xvi
Ceratostigma Griffithii, 18, 107
Cheilanthes farinosa, xvi, 198
Cherry (tree), 110, 134
Cimicifuga foetens, 172
Clematis, 18, 76, 81, 99
Clematis acuminata, 161
Clematis connata, 107
Cochlearia scapiflora, 112
Compositae, xv, 75
Corydalis, 112
Corylus ferox, 179
Crataegus, xvi
Cremanthodium, 74, 75, 79, 111, 138, 143, 199
Cremanthodium Thomsoni, xvi
Crepis, 199
Cruciferae, xv
Cyananthus, 184
Cyananthus lobatus, 193
Cymbidium, 161
Cynoglossum, 76
Cypress, 138
Cypriedium. tibeticum, 129
Cypripedium luteum, 129
Cypripedium Wardii, 193
Daisy (Ox-eye), 89
Delphinium, 79
Delphinium Brunonianum, 143
Deutzia, 18, 128
Didissandra lanuginosa, xvi, 107
Dogwood, 128

INDEX

Draba alpina, 112, 143
Dracocephalum, 112, 140, 149, 199
Dracocephalum tanguticum, 85
Dragonheads, 149
Dregea sinensis, 198
Eremurus chinensis, 198
Euonymus, 164
Fir, 34, 46, 65, 66, 89, 106, 108, 114, 119, 161, 167, 172, 185
Fir, (Silver), 62, 173, 180, 194
Fraxinus floribunda, 56
Fritillaria, 196
Gamblea ciliata, 53
Gaultheria codonantha, 193
Gentian (Lhagu), 155, 174
Gentiana, xv
Gentiana barbata, 76
Gentiana Georgei, 118, 121
Gentiana gilvostriata, 193
Gentiana sino-ornata, 118, 121, 125
Gentiana trichotoma, 125, 193, 199
Gentiana Wardii, 125
Geranium, 76, 90, 196, 199
Globe flower, 66, 86, 113
Hawthorn, 107
Hazel, 179
Hippophäe, 78, 89, 90
Holboellia, 18
Hollies, 171, 194
Honeysuckle, 61, 67, 128
Hornbeam, 29, 194
Horse Chestnut, 193
Ilex, xv, 164
Incarvillea, xv
Incarvillea brevipes, 81
Indigofera, 18, 107
Iris, 20, 26, 30, 52, 58, 86, 159, 169
Iris (bog), 90
Iris Clarkei, 195
Iris Delavayi, 195
Iris goniocarpa, 195
Iris kumaonensis, 195
Iris Wattii, xv, 195
Jasmine, 52, 107, 180
Juniper, xv, xvi, 53, 67, 75, 76, 81, 89, 110, 139

Labiatae, xv
Lactuca, 199
Lagotis, 112
Larch, 34, 53, 165, 171
Larix Griffithii, 62
Larkspur, 143
Laurels, 194
Leycesteria crocothyrsos, 193
Lilac, 180
Lilium giganteum, 53, 164
Lilium Wardii, xv, 30, 49, 60, 66, 67, 68, 128, 159, 193
Litsaea, 18, 34, 35, 128, 159
Lonicera, xv, 75, 89, 157
Lonicera hispida, 157
Lonicera Webbiana, 81
Loranthus odoratus, 9
Macaranga, 18
Magnolia, 20, 26, 175, 194
Magnolia Campbellii, 28, 187
Magnolia globosa, 53
Mahonia calamicaulis, 20
Maple, 20, 26, 35, 53, 54, 164, 167, 179, 180, 194
Martagon Lily, 49, 52
Meconopsis Baileyi, 193
Meconopsis betonicifolia, xvi, 172
Meconopsis horridula, 64, 81, 112
Meconopsis integrifolia, 172
Mentha, xvi, 107
Michelia lanuginosa, 175
Molineria oligantha, 193
Monkshood, 90, 113, 154, 155
Morina, 85
Nomocharis, xv, 30, 49, 67, 139, 173, 174, 185, 196
Nomocharis nana, 68
Nomocharis pardanthina, 68, 174
Nomocharis Souliei, xvi, 64, 68, 174
oak, 18, 20, 134, 164, 179, 194
Omphalogramma Forrestii, xvi
Omphalogramma Souliei, 185
Onion, 185
Onosma Hookeri, xvi, 129
Orange, mock (Philadelphus), 128

INDEX

orchid, 31, 52, 86, 129
Orchis, 66
Osyris arborea, 9
Oxytropis, xv, 66, 74, 196
Paeony, 134
Paeony Delavayi, 134
Paraquilegia microphylla, xv, 66, 69, 86, 96, 123
Parnassia nubicola, 90
Pedicularis, xv, 76, 79, 89, 199
Philadelphus, 18
Picea lichiangensis, xv, 110, 154, 196
Picea Morinda, xv
Pine, 11, 19, 20, 21, 25, 28, 30, 34, 49, 60, 171, 175, 178, 179, 194
Pinus excelsa, xv, 199
Pinus Khasia, 18
Pittosporum floribundum, 18
Plane tree, 171
Podophyllum versipelle, 53
Polygonum, xv, 74, 112, 199
Poplar, 61, 110, 134, 196
Potentilla, 74, 96, 112
Potentilla fruticosa, xv, 75
Primula, xv, 64, 66, 67, 74, 76, 86, 87, 96, 110, 116, 121, 122, 143, 172, 173, 197
Primula ('Nivalis'), 123
Primula Agleniana, xvi
Primula alta, 196
Primula Bulleyana, 193
Primula calliantha, 197
Primula chungensis, 53, 193
Primula Clutterbuckii, 193
Primula concholoba, 193
Primula cyanantha, 74
Primula denticulata, 64
Primula Dubernardiana, 110
Primula Florindae, 193
Primula geranioides, 185
Primula macrocarpa, 111
Primula macrophylla, 143
Primula melanops, 197
Primula microdonta, 193
Primula minor, 112, 199
Primula Normaniana, 193
Primula rubra, 193
Primula sikkimensis, 58, 66, 67, 79, 125, 138
Primula szechuanica, 74, 154
Primula tibetica, 79, 89, 121
Prunus Padus, 18
Ranunculaceae, xv
Ranunculus, 74, 112
Rhododendron, xv, 20, 26, 30, 48, 53, 58, 59, 62, 64, 65, 67, 70, 74, 76, 92, 106, 108, 116, 125, 128, 139, 140, 149, 154, 156, 164, 167, 171, 172, 173, 179, 180, 183, 185, 187, 194, 196, 199
Rhododendron arboreum, 28
Rhododendron Beesianum, 59, 62, 65, 134
Rhododendron bullatum, 30, 31, 34, 161, 164
Rhododendron cerasinum, 58, 184
Rhododendron cinnabarinum, 58, 59
Rhododendron crebreflorum, 74, 193
Rhododendron fulvoides, 59, 164
Rhododendron hylaeum, 34
Rhododendron imperator, 193
Rhododendron keleticum, 75
Rhododendron kongboense, 59
Rhododendron lanatum, 180, 184, 200
Rhododendron lepidotum, 128
Rhododendron leucaspis, 193
Rhododendron lysolepis, 193
Rhododendron manipurense, 161, 164
Rhododendron megacalyx, 30
Rhododendron melinanthum, 193
Rhododendron mishmiense, 193
Rhododendron patulum, 193
Rhododendron pruniflorum, 58
Rhododendron repens, 68
Rhododendron riparium, 74, 193
Rhododendron sanguineum, 59, 69, 185
Rhododendron sigillatum, 74, 76, 108, 154
Rhododendron sinogrande, xv, 34
Rhododendron tanastylum, 164
Rhododendron telmateium, 74, 108

INDEX

Rhododendron Thomsonii, 128
Rhododendron trichocladum, 58
Rhododendron tsarongense, 108, 128
Rhododendron vellereum, 154
Rhododendron vesiculiferum, 164
Rhododendron virgatum, 164
Rhododendron Wardii, 62, 65, 193
Rhus, 157
Rosa bracteata, 18
Rosa sericea, xv, 81, 107
Rowan, 180, 182
Rubus lasiocarpus, 18
Rubus moluccanus, 18
Salix, xvi
Salvia, 75, 158
Saussurea, xv, 140, 143, 196
Saussurea gossypifera, 112
Saxifraga, xv, 110, 112, 140
Saxifraga candelabrum, 198
Schefflera shweliensis, 29
Sedum, xvi, 143
Selaginella involvens, xvi, 107
Senecio, 75

Sophora, 105
Sophora viciifolia, xvi, 107
Sorbus, xv
Spindle tree, 164
Spiraea, 75
St. John's Wort, 185
Stellera chamaejasme, xv, 30, 52, 79, 81
Strobilanthes, 164
Swertia, 199
Syringa, xv
Tamarisk, 61, 74
Trollius, 74, 76, 86, 90, 113, 125, 149
Tsuga, 53
Tsuga yunnanensis, xv, 34
Tupidanthus calyptratus, 18
Vaccinium glauco-album, 164
Viburnum, 18, 34, 128
Viburnum Wardii, 34
Violets, 35, 52
Wikstroemia, xvi, 107
Willow, 18, 29, 74, 128
Zanthoxylum, 18

INDEX

II
PLACES

Aju Chu, 109
Arig, 151
Assam, xv, 8, 13, 23, 37, 47, 52, 56, 91, 108, 157, 173, 174, 177, 182, 183, 184, 185, 188
Ata, 34, 46, 51
Ata (glacier), 71, 76, 82, 155, 196
Ata (river), 49, 50, 51, 73, 161
Ata Kang La, 53, 54, 56, 71, 75, 82, 84, 85, 93, 98, 120, 125, 128, 134, 135, 148, 149, 151, 152, 154, 196, 199
Burma, xi, xiii, 9, 18, 72, 108
Bütang, 82, 84, 85, 89, 91, 96, 114
Calcutta, 2, 159
Cha Che, 187, 188
Chamdo, 4, 9, 15, 19, 23, 47, 51, 94, 100, 118, 121, 126, 138, 151, 195, 198
Changra, 151
Cheti La, 65, 67, 68, 71, 73, 156
Chibaon, 190
Chimdru (river), 162, 166, 199
Chömbö, 53, 54, 70, 71, 200
Chömbö (glacier), 71
Chutong, 59, 61, 62, 64, 65, 67, 68, 69, 70, 72, 76, 156, 157, 196
Delei (river), 184, 185, 187, 189
Delei (valley), 174, 176, 177, 182, 188, 190, 200
Denning, 4, 13, 191, 193, 194
Dibang (river), 32, 164, 170, 193, 200
Dibang (valley), 2, 173
Dihang (valley), 2, 32, 162
Dorge Tsengen, 78, 82, 84
Draya, 162, 168
Draya Gompa, 116, 134
Dreyi, 5, 191
Dri, 38, 41, 46, 164, 175, 176, 177, 180, 186, 187, 188, 195, 200
Dri (river), 200
Dri La (valley), 177, 200
Dri Pass, 174, 184
Dza, 105, 106

Gechi Gompa, 51, 60, 196
Giwang, 13, 16, 20, 24, 25, 26, 32, 35, 121, 175, 177, 195
Gompo Ne, 119
Gongsar Gompa, 100, 102, 103
Gyantse, xii, 4, 137
Jara La, 172, 173
Kailas (Mt.), 199
Ka-Kar-Po, 138, 199, 200
Kam, 13, 18, 23, 93
Kangri Karpo La, 159, 161, 162, 166, 200
Kaso, 187
Kongbo Tsangpo, 119, 135
Lhagu, 76, 78, 100, 135. 136, 152, 154, 155
Lhagu (glacier), 78, 92
Lhagu (lake), 76, 135
Lhagu (valley), 156
Lhagu Chu, 78
Lhasa, xii, 4, 12, 15, 23, 51, 81, 86, 88, 93, 100, 103, 115, 118, 137, 152, 195, 199
Lho Dzong, 81, 100, 137
Lho-Tsa-Ri, 199
Lohit (river), xiv, 2, 5, 11, 14, 19, 126, 148, 191, 193, 195
Lohit (valley), xv, 2, 5, 6, 8, 18, 19, 52, 139, 177, 180, 190, 194
Minutang, 189
Minzong, 7, 9, 10
Mishmi Hills, xvi, 4, 5, 9, 13, 16, 30, 31, 108, 171, 193, 200
Modung, 14, 51, 60, 67, 72, 148, 154, 157, 159, 160, 169, 195
Mugu, 31, 170, 171, 172, 174, 175, 195
Nagong (district), 93, 94, 124
Nagong (river), 73, 84, 93, 98, 119, 120, 135
Nagong Chu, 200
Nagong Chu (river), 118, 119, 120, 134

INDEX

Namcha Barwa, xiii, xiv
Ningri Tangor, 89, 96, 115, 116, 118, 119, 121, 122, 123, 149
Pangum, 6
Parangkon, 182
Pashu, 94, 100, 138
Pemako, 159, 166
Pö Yu La, 84
Pomé, 86, 119, 120, 148, 166
Po-Tsangpo, 119
Poyü Tso (lake), 98, 148
Poyul, 120
Puti, 106, 107, 109
Putsang, 160, 161, 162, 164, 166, 168, 178, 195, 197, 200
Rima, xiv, xv, 1, 2, 5, 6, 7, 9, 10, 11, 13, 14, 15, 16, 18, 19, 20, 22, 23, 30, 37, 38, 46, 47, 48, 51, 54, 56, 61, 72, 84, 100, 115, 118, 126, 134, 138, 139, 162, 169, 177, 178, 191, 194, 195, 200
Rong Tö (valley), 14, 16, 29, 32, 37, 48, 51, 125, 128, 134, 152, 161, 162, 170, 175, 180, 195, 199, 200
Rong Tö Chu (river), 14, 19, 50, 51, 98, 134, 148, 159, 160, 161, 164, 171, 195, 197, 198, 200
Rongyul, 36, 46, 48, 49, 161, 168, 195
Ru, 111, 140, 142, 143, 144, 145, 148
Ru La, 142
Ruowa, 119
Sachung, 195
Sadiya, 2, 4, 5, 7, 14, 15, 19, 24, 32, 33, 37, 51, 60, 139, 159, 160, 171, 177, 179, 188, 191, 192
Saikhoa Ghat, 2
Salween, xiii, xiv, xv, 1, 81, 84, 86, 87, 89, 96, 97, 98, 99, 100, 104, 105, 106, 107, 108, 110, 128, 134, 139, 197, 198, 199
Salween (valley), xvi, 108, 197, 198
Sangachu Dzong, 47, 82, 110, 115, 118, 119, 121, 124, 125, 126, 137, 138, 148, 149, 197, 198, 200
Shigatang, 22, 194
Shingki Dzong, 166
Shinki Gompa, 199
Shoshi Dzong, 98, 135, 139, 145, 147, 148
Showa, 119, 120, 135
Shugden (lake), 78, 93, 135
Shugden Gompa, 1, 3, 4, 47, 54, 59, 68, 72, 78, 79, 81, 82, 84, 85, 86, 88, 90, 91, 92, 93, 96, 98, 100, 104, 108, 109, 110, 114, 115, 120, 123, 126, 133, 135, 136, 137, 138, 139, 148, 149, 152, 157, 159, 195, 196, 197, 198, 199, 200
Shukdam, 62
Solé, 16, 20, 29, 30, 31, 32, 33, 37, 46, 68, 168, 169, 195
Suku, 128, 160
Suku (river), 51
Szechuan, 15, 136, 161, 162
Tajabum, 177, 186, 187, 188
Tang La, 196
Tangon (river), 31, 32, 173
Theronliang, 5, 6, 191
Tidding (river), 5, 191
Tö Kang-Ri, 199
Töyul, 175, 195
Traki La, 84, 93, 111, 134, 140, 144, 199
Trashitze Dzong, 99, 100
Tsa Chu (river), 100, 109
Tsangpo, xiii, xiv, 2, 71, 86, 97, 107, 108, 195, 197, 199
Tsarong, 104
Tsiling Chu, 126, 134
Tzengu Chu, 92, 139
Watak, 104, 105
Yangoong, 151
Yangtse Kiang, xiii, xiv, 136
Yatsa, 78, 135, 152
Yatsa Chu, 124
Yindru, 110, 113
Zayul, xiv, 1, 2, 9, 10, 11, 16, 21, 22, 23, 25, 32, 40, 47
Zayul (Governor of), 12, 14, 15, 37, 110
Zayul Chu, 126, 134, 195
Zayul Ngu Chu, 50, 164, 168, 197
Zigar, 105, 109, 110
Zo La, 125, 126, 134, 149, 150, 198, 199, 200
Zo La Chu (river), 78

INDEX

III
PERSONS

A. K. (Rai Bahadur Kishen Singh Milamwal), 1, 82, 100, 196
Bailey, Lt.-Colonel F. M., 1, 82, 89, 199
Brooks-Carrington, 1, 3, 6, 12, 13, 26, 28, 46, 61, 66, 67, 69, 70, 73, 79, 121
Cawdor, Lord, 68, 135
Chimi (servant), 121, 130, 131, 132, 146, 149, 179, 188
Clutterbuck, Hugh, 177, 178, 188, 189, 200
Jaglum, 5, 6, 7, 8
Kalon Lama, 51, 195
Kaulback, R., 1
Kele, 121, 130, 131, 132, 133, 139, 146, 149, 151, 152
Kyipu, 14, 51, 149, 161
Lobsang, 41, 42, 43, 47
Nimnoo, 4, 5, 6
Pinzo (cook), 3, 68, 73, 131, 137
Rockhill, W. W., 198
Tashi Tendu, 3, 13, 73, 96, 116, 121, 122, 130, 131, 132, 133, 139, 140, 142, 149, 152, 156, 177, 191
Teichman, Eric, 198
Tsumbi (Sirdar), 2, 3, 7, 11, 12, 14, 19, 24, 25, 35, 37, 38, 40, 41, 42, 59, 60, 61, 66, 67, 68, 73, 78, 79, 80, 86, 87, 88, 99, 110, 116, 118, 124, 130, 131, 132, 133, 137, 139, 140, 147, 152, 156, 158, 160, 169, 178, 179, 186, 188
Wunju, 13, 24, 32, 177, 178, 179, 186, 187, 188, 189, 190

IV
ANIMALS, BIRDS, ETC.

Babblers, 65, 81, 92, 138, 190
Bear (Ursus torquatus), 33, 174
Beetle, Oil, 109
Bugs, 109
Butterflies, 20, 91, 108, 109, 198
Choughs, 75, 92, 138
Coppersmith, 190
Cuckoo, 28, 65
Eagle, White-Tailed, 92
Earwigs, 59
Finch, Rose, 65, 92
Flycatcher, 65
Frog (Bujo viridis), 108
Gasteropods (Limnaea), 89
Gazelle, Tibetan, 142
Gooral, 32, 33, 162
Grasshoppers, 91, 108
Hare, 75, 81, 91, 149
Hare, Pigmy (Ochotona), 65, 92
Hoopoe, 92
Larks, 155
Leopard, Snow, 103
Lizard, 108, 198
Magpie, 75, 92
Marmot, 75, 92
Marten, 33
Monal (Lophophorus sclateri), 56, 65, 67
Musk deer, 162
Nightjar, 62
Parakeet, 28
Partridge, 92
Pigeon, 103
Pigeon, Rock, 92
Pigeon, Snow (Columba leuconota), 65
Robin, 65
Sandflies, 54
Slugs, 90, 139
Snakes, 49
Snipe, Jack, 92
Snipe, Solitary, 138
Squirrel, 33
Takin (Budorcas taxicolor), 32, 33, 162, 166, 167
Ticks, 165
Toad, 91, 198
Voles, 65
Woodpecker, 120

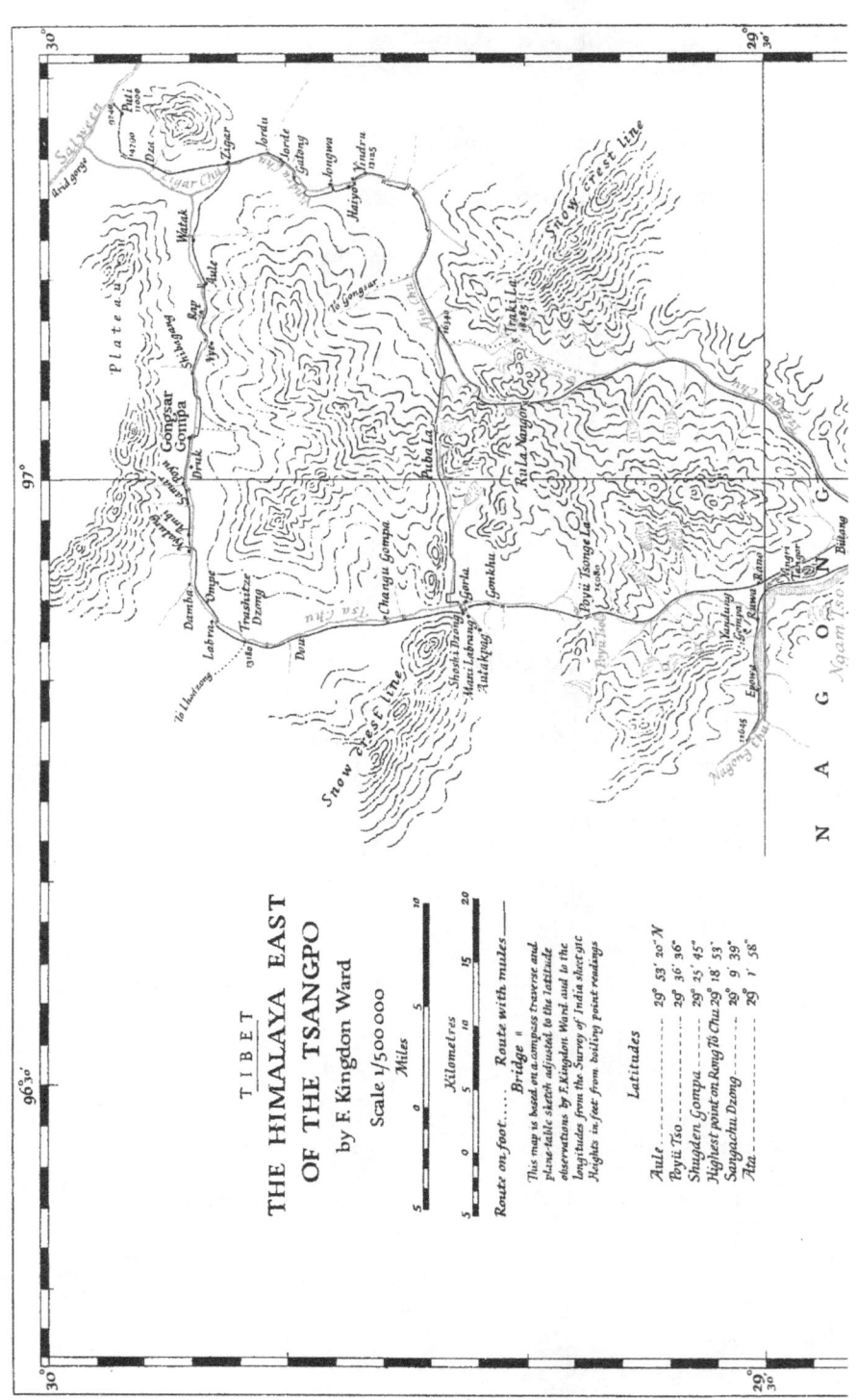

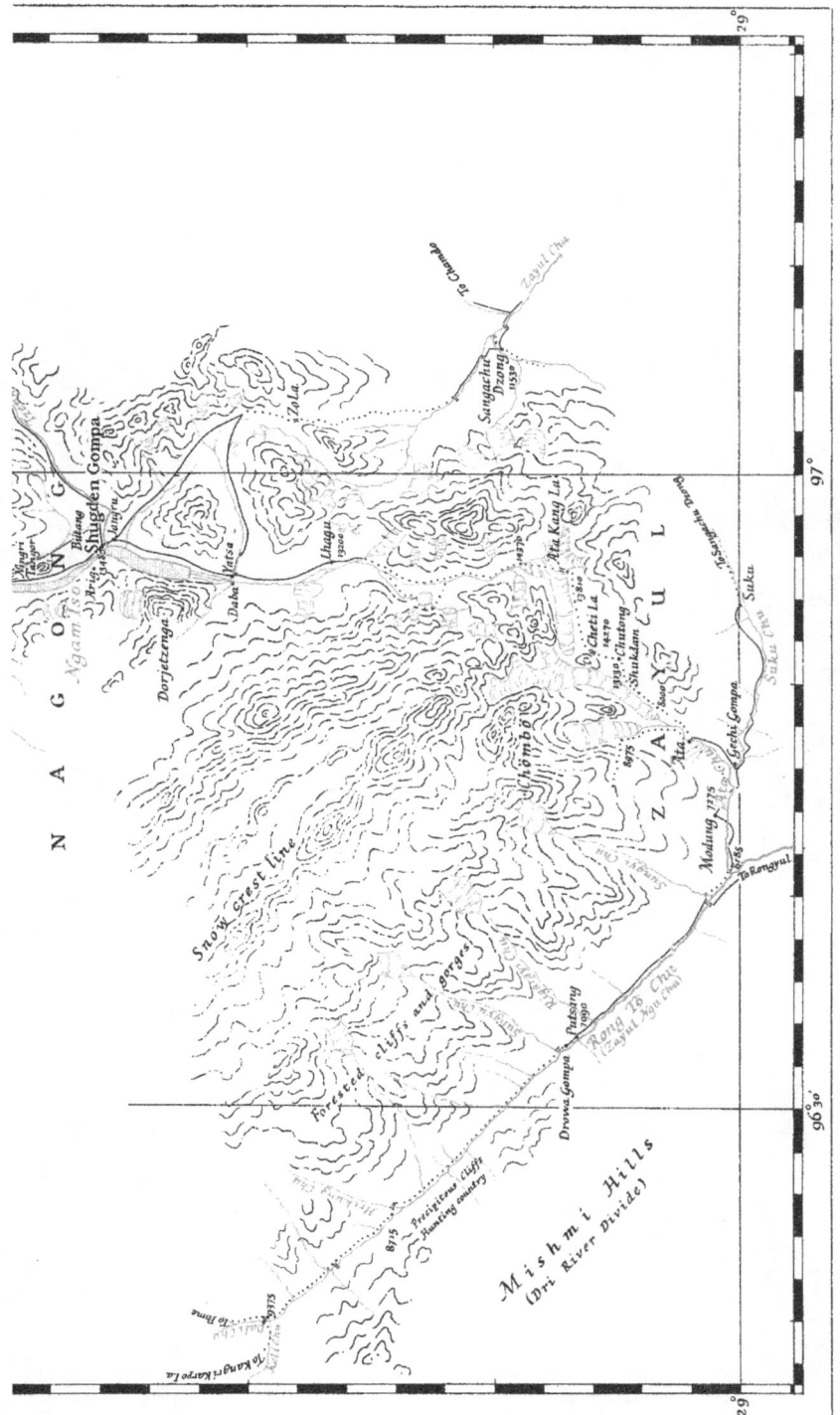

Drawn at the Royal Geographical Society and reproduced by permission.

ABOUT THE AUTHOR

Frank Kingdon-Ward (1885–1958), OBE, the son of a leading British research botanist and University of Cambridge professor, graduated with honours from Cambridge in Natural Sciences, in 1906, and subsequently accepted a teaching position in Shanghai. In 1909, he broke this contract to travel across central and western China; while the expedition was not specifically botanical in purpose, Kingdon-Ward assembled a small collection of plant samples enroute, which were subsequently presented to the Botany School of Cambridge.

In 1911, he was engaged by the horticultural firm of Bees Ltd. of Liverpool, to collect botanical specimens in Yunnan, an assignment which began for Kingdon-Ward a lifelong career as a professional explorer and plant collector.

In all Kingdon-Ward made a total of twenty-two expeditions, spanning a period of some forty-five years, in western China, northern Burma, Assam and south- eastern Tibet; much of this travel involved extreme hardship, and was undertaken at great risk to his health and personal safety. On all of his travels, Kingdon-Ward not only collected rare and valuable plants, but also surveyed and mapped previously uncharted territories.

During WWI, Kingdon-Ward served in the Indian Army, achieving the rank of captain. He was in Burma when Japanese forces invaded that country in WWII, but with his intimate knowledge of the terrain, he was successful to evade capture and enter India, where he was subsequently engaged to instruct British forces on techniques for jungle survival.

But it was Kingdon-Ward's contribution to botany which remains his foremost legacy, as the plant genera *Kingdon-wardia*, and *Wardaster*, and the wealth of previously unknown plant species now bearing the suffixes *wardii* and *kingdonwardii* attest. In the course of his long and productive career, Kingdon-Ward received honours for his botanical contributions from many learned societies, including the Royal Geographic Society, the Royal Horticultural Society, the Linnean Society of London and the Royal Central Asian Society, among others.

Frank Kingdon-Ward was also a prolific writer, describing his travels in 25 books, as well as in over 700 articles in both scholarly and popular botanical journals. In addition to the present volume, Orchid Press has published the following titles, long out of print:

> *In Farthest Burma* (first published 1921)
> *Burma's Icy Mountains* (first published 1949)
> *Return to the Irrawaddy* (first published 1956)

www.ingramcontent.com/pod-product-compliance
Lightning Source LLC
Chambersburg PA
CBHW020837160426
43192CB00007B/688